Social Unrest and American Military Bases in Turkey and Germany since 1945

Over the past century, the United States has created a global network of military bases. While the force structure offers protection to U.S. allies, it maintains the threat of violence toward others, both creating and undermining security. Amy Austin Holmes argues that the relationship between the U.S. military presence and the non-U.S. citizens under its security umbrella is inherently contradictory. She suggests that, although the host population may be fully enfranchised citizens of their own government, they are at the same time disenfranchised under the U.S. presence. This study introduces the concept of the "protectariat" as they are defined not by their relationship to the means of production, but instead by their relationship to the means of violence. Focusing on Germany and Turkey, Holmes finds remarkable parallels in the types of social protest that occurred in both countries, particularly nonviolent civil disobedience, labor strikes of base workers, violent attacks and kidnappings, and opposition parties in the parliaments.

Amy Austin Holmes is Assistant Professor of Sociology at the American University in Cairo and Postdoctoral Fellow of International Studies at Brown University's Watson Institute. She received her Ph.D. from Johns Hopkins University and an M.A. from the Freie Universität Berlin. Her work has been published in *Mobilization*, *South Atlantic Quarterly*, the *Cairo Review of Global Affairs*, the *Baltimore Sun*, *Ahram Online*, and the *Atlantic Council*. Her research interests include social movements, revolutions, critical security studies, and U.S. foreign policy in Europe and the Middle East.

Social Unrest and American Military Bases in Turkey and Germany since 1945

AMY AUSTIN HOLMES

*American University in Cairo and
Brown University*

CAMBRIDGE
UNIVERSITY PRESS

CAMBRIDGE
UNIVERSITY PRESS

32 Avenue of the Americas, New York, NY 10013-2473, USA

Cambridge University Press is part of the University of Cambridge.

It furthers the University's mission by disseminating knowledge in the pursuit of education, learning, and research at the highest international levels of excellence.

www.cambridge.org
Information on this title: www.cambridge.org/9781107019133

First published 2014

A catalog record for this publication is available from the British Library.

Library of Congress Cataloging in Publication Data
Holmes, Amy Austin, 1973–
Social unrest and American military bases in Turkey and
Germany since 1945 / Amy Austin Holmes.
   pages cm.
Includes bibliographical references and index.
ISBN 978-1-107-01913-3 (hardback)
1. Military bases, American – Germany.   2. Protest movements – Germany –
History.   3. Sociology, Military – Germany.   4. United States – Relations –
Germany.   5. Germany – Relations – United States.   6. Air bases, American – Turkey.
7. Protest movements – Turkey – History.   8. Sociology, Military – Turkey.
9. United States – Relations – Turkey.   10. Turkey – Relations – United States.   I. Title.
UA26.G3.H65   2013
355.70943–dc23   2013035747

ISBN 978-1-107-01913-3 Hardback

Contents

Acknowledgments

The debts I have incurred in writing this book may never be repaid. A word of thanks is in order. As far back as I can remember, I have always admired scholars who have the ability to paint the broad contours of large-scale societal change with a Breughelian attention to detail. Beverly Silver and the late Giovanni Arrighi, my Ph.D. advisers at Johns Hopkins University, have – in my mind at least – pretty near mastered the art. Readers familiar with their work may see their influence on my own research, but I have not attempted the same. This is not a study of the longue durée, but rather of the middle range. I am also indebted to the continuing support of Joel Andreas, as well as Margaret Keck and Siba Grovogui.

Of course, a book requires more than inspiration to be completed, but also a variety of financial and practical resources, particularly when the fieldwork is conducted in multiple countries. The research for this project was only possible with the generous support of the Fulbright Foundation, the Institute of Turkish Studies, the American Institute of Contemporary German Studies, the American Council on Germany, the Transnational Institute, and the American University in Cairo. I would like to thank Jan Philipp Reemtsma and Bernd Greiner at the Hamburger Institut für Sozialforschung, where I spent a year as a Fulbright Fellow. A number of institutions supported shorter research trips. For their support, I would like to thank Klaus Dörre and the entire Kolleg Post-Growth Societies at the Friedrich-Schiller-Universität Jena, Paul Seabright at the Institute of Advanced Studies in Toulouse, and Vivek Chibber and Marilyn Young at New York University. During various research trips to Istanbul, I was able to use the library and other resources of the American Research Institute in Turkey, the Swedish Research Institute, and the French Institute of Anatolian Studies. Finally, I would like to thank Lisa Anderson, Amr Shaarawi, Graham Harman, and Nathaniel Bowditch at the American University in Cairo, as AUC provided me with a research grant and a semester-long sabbatical leave to complete the book manuscript.

Obtaining documents from U.S. government archives became more difficult after 2003. I would like to thank all of the archivists who, in the face of slashed budgets and increased demands on their time, were still able to help me locate valuable materials. In particular I would like to thank Richard Boylan at the National Archives and Records Administration (NARA). Ellen Keith at the Johns Hopkins library was also always reliable and very resourceful. Last but not least, for believing my manuscript should be published, I owe many thanks to Lew Bateman at Cambridge University Press.

Over the course of my research I conducted more than 100 semi-structured interviews. I have listed their names in the bibliography and I would like to thank each and every one of them for dedicating their time to this project. In particular, I would like to thank Otfried Nassauer, Erich Schmidt-Eenboom, Marianne Zepp, and Ayşe Gül Altınay for sharing their expertise. I would also like to thank James Blaker, Steven Szabo, Kent Calder, Jackson Janes, Michael Hanpeter, and Benton Moeller for their insights and for helping me to locate important documents.

The late Charles Tilly invited me to present two chapters of the dissertation at his weekly seminar at Columbia University, and I would like to thank all the participants in the workshop for their feedback, in particular James Jasper, John Krinsky, Christian Davenport, Nelly Lahoud, Sun-chul Kim, and Roy Licklider. Toward the crucial final phase of writing and revising, Yahya Madra was there with his style sheet expertise and Peter Hruschka with his hawk eye for editing and invincible you-can-do-it mentality. A number of scholars working on similar issues have been exceptionally generous with their time and have encouraged me to disregard what only seem to be insurmountable obstacles and stay the course: the late Chalmers Johnson, Catherine Lutz, Joseph Gerson, David Vine, and Emira Woods. A long list of other friends and colleagues have listened patiently to my military base–related ruminations at various way stations during my triangular travels between Baltimore, the German hinterland, and tea gardens along the Bosphorus and the Nile, including: Conny Tüxsen, Gülru Çakmak, Frank Deppe, Guido Speckmann, Wilbert van der Zeijden, Deniz Yükseker, Çağla Diner, Evrim Binbaş, Christian Frings, Conny Weissbach, Jake Lowinger, Kevan Harris, Bilgin Ayata, Felipe Hough, Ladan Akbarnia, Nicole Aschoff, Michael McCarthy, Alex Stickler, Stefan Schmalz, David Salomon, Necmi Bayram, Anja Hälg, Alessandro Pelizzari, Renaud Martel, Izabela Buraczewska, Ertan Keskinsoy, and the late Ulus Baker.

Finally, this book is dedicated to the unsinkable ship that is my family. To my brother Doug and his daughter Rebekah. And to my parents, Ann and Bert Holmes, who, combining just the right amount of patience with impatience, always believed I would finish.

List of Acronyms

AKP	Adalet ve Kalkinma Partisi	Justice and Development Party
AP	Adalet Partisi	Justice Party
AWACS	Airborne Warning and Control System	
BRD/FRG	Bundesrepublik Deutschland	Federal Republic of Germany
CDU	Christlich Demokratische Union	Christian Democratic Union
CENTO	Central Treaty Organization	
CHP	Cumhuriyet Halk Partisi	Republican People's Party
DCSHNA	Deputy Chief of Staff of Host Nation Activities	
Dev-Genç	Devrimci Gençlik	Revolutionary Youth
DGB	Deutsche Gewerkschaftsbund	Confederation of German Trade Unions
DİSK	Devrimci İşçi Sendikaları Konfederasyonu	Confederation of Revolutionary Labor Unions
DoD	Department of Defense	
DP	Demokrat Parti	Democrat Party
EUCOM	European Command	
FDP	Freie Demokratische Partei	Free Democratic Party
Harb- İş	Türkiye Harb Sanayi ve Yardimci Iskollari Isçileri Sendikasi	Federation of Defense Industry and Allied Workers Union
IGPBS	Integrated Global Presence and Basing Strategy	
INF	Intermediate Range Nuclear Forces	

JUSMAT	Joint U.S. Military Mission for Aid to Turkey	
METU	Middle East Technical University	
MRP	Master Restationing Plan	
NARA	National Archives and Records Administration	
NATO	North Atlantic Treaty Organization	
NDR/MDD	Milli Demokratik Devrim	National Democratic Revolution
NSC	National Security Council	
OBC	Overseas Basing Commission	
ÖTV	Öffentliche Dienste, Transport und Verkehr	Public Service and Transportation Union
PDS	Partei des Demokratischen Sozialismus	Party of Democratic Socialism
PKK	Partiya Karkerên Kurdistan	Kurdistan Workers' Party
PX	Post Exchange	
QDR	Quadrennial Defense Review	
RAF	Rote Armee Fraktion	Red Army Faction
REFORGER	Return of Forces to Germany	
RZ	Revolutionäre Zellen	Revolutionary Cells
SACEUR	Supreme Allied Commander Europe	
SOFA	Status of Forces Agreement	
SPD	Sozialdemokratische Partei Deutschlands	
THKO	Türkiye Halk Kurtuluş Ordusu	Turkish People's Liberation Army
THKP-C	Türkiye Halk Kurtuluş Partisi-Cephesi	Turkish People's Liberation Front
TIP	Türkiye İşçi Partisi	Turkish Labor Party
Türk-İş	Türkiye Işçi Sendikaları Konfederasyonu	Confederation of Turkish Trade Unions
TUSLOG	Turkish U.S. Logistics Group	
USAREUR	U.S. Army Europe	
USFLO	U.S. Forces Liaison Office	
WHNS	Wartime Host Nation Support	

I

Introduction

The Global American Military Presence in Comparative Perspective

INTRODUCTION

As empires rise and fall, they expand and contract in space. From the Roman Empire to the British Raj, this process has involved the conquest and acquisition of territories and their populations during periods of expansion, and their subsequent loss during periods of decline. This process, however, was not discontinued in the postcolonial era. Territorial acquisition – and the contention it calls forth – is still at the heart of great power politics; it has merely changed in form. The westward expansion of the American frontier and the creation of a continental-sized nation-state have been documented by historians who have underlined how territorial acquisition was constitutive of what would become the United States of America. And yet, American expansion took another form as well. Beginning in 1898, the United States began to acquire overseas military bases, assembling a fragmented base structure out of the "leftovers" of the European empires. By 1938, there were fourteen American military bases outside the continental United States. A decade later, the U.S. overseas basing network included several thousand facilities and was larger than any of its European counterparts. As there is no single office in the U.S. government that is in charge of the issue, it is entirely possible that no one ever knows exactly how many overseas military bases the United States owns, operates, or has access to at any given time.[1] And yet there is general agreement that the

---

[1] Giving a precise estimate of the number of overseas bases is difficult, if not impossible, for two additional reasons. First, there is little agreement over what "counts" as a base, and even official definitions have changed over time. Second, the secretive nature of some facilities or agreements means that some data are simply not disclosed to the public. Even the annual Base Structure Report published by the Department of Defense, which is the most comprehensive source of information, does not include certain categories of facilities. For varying estimates, see Harkavy 1982, Blaker 1990, and Johnson 2010.

overseas network of bases has expanded at three important historical junctures: (1) after the Spanish-American War in 1898, (2) during and after World
War II, and (3) after the end of the Cold War.

Indeed, the creation of this vast network of overseas bases is one of the most
striking testaments to the transformation of the global landscape during the
middle of the twentieth century. As European empires were contracting and
shrinking back to the size of smallish nation-states, the American empire was
expanding – and in a rather unprecedented fashion.

In explaining the emergence of the United States as a global power, scholars
have often focused on its role in the global economy, its military prowess, and
on its various manifestations of soft power. In doing so, much has been done
to document the complex and multifaceted nature of U.S. hegemony, and yet
they have often overlooked what is particularly unique about the American
empire. After all, other great powers have risen through the dynamism of their
economies, the strength of their militaries, or the dissemination of their cultures. None of them, however, stationed their soldiers and military outposts
in hundreds of sovereign states around the world on a permanent, open-ended
basis. And yet this is precisely what the United States has done for more than a
century, regardless of whether or not it has been engaged in combat operations.
Although previous imperial powers deployed their troops to colonial possessions, no other hegemonic power was able to convince sovereign states to play
"host" to an enduring military presence during peacetime.

The current period of American hegemony is unusual for a second reason as
well. From the fifteenth right up to the early twentieth century, the era of colonialism was defined by competition as European states competed with each
other, indeed scrambled, to acquire overseas territories. During the Cold War,
competition between the United States and the Soviet Union not only took the
form of an arms race, but also of a "base race." Geopolitik was resurrected as
great power competition to acquire access to military bases and port facilities
in foreign countries seemed to bear out older theories of heartland versus rimland strategies.[2] Since the collapse of the USSR, the United States has not only
continued to maintain but has even expanded its network of military bases,
penetrating Eastern Europe, Central Asia, the Middle East, and Africa. This
empire of overseas bases has no equal or rival. It is now characterized by a near-
absence of competition, or what was previously referred to as inter-imperialist
rivalry. Although fierce competition persists in the economic sphere, it is no
longer the driving force behind foreign base acquisitions. Ironically, just as the
United States had acquired territorial access rights in more regions of the world
than ever before, it became fashionable to speak of the "deterritorializing"

---

[2] The terms "heartland" and "rimland" or "marginal crescent" can be traced to Halford
Mackinder's 1904 book *The Geographical Pivot of History*, which was later criticized by Alfred
Mahan who argued in favor of the historical primacy of sea power versus land power. Both,
however, saw the Eurasian landmass as the key region in global politics.

nature of American power.[3] The means by which overseas territory is obtained may have changed, but the ends – control over or access to foreign soil – has not. Instead of a colonial empire, the United States operates a baseworld.

Between 1950 and 2005, no less than 170 countries have hosted some form of an American military presence.[4] These facilities range in size from bare-bones landing strips that may go unnoticed by outsiders to gargantuan military base complexes sprawling into civilian communities. Yet despite their centrality, the role of American overseas military bases in projecting U.S. power abroad, and therefore forming the very underpinning on which U.S. hegemony was built and maintained, has remained a relatively unexplored frontier within the academic community. The most quintessential aspect of the American empire is also the most understudied.

Not only is the size of this base structure unprecedented, but so is its modus operandi. Whereas previous hegemonic powers such as Great Britain created bases in colonial possessions by virtue of its sovereign control, the postwar growth of the American basing network went hand-in-hand with the postwar decolonization process and the expansion of the Westphalian system that resulted in a veritable explosion of the number of sovereign states. The proliferation of sovereignty has transformed both individual states as well as the structural landscape of international politics. Historically, empires developed different ways of managing difference within their realms. Especially European colonial empires in the nineteenth and twentieth centuries tended to create binary distinctions between the colonizers and the colonized, the metropole and the periphery, the viceroys and their subjects.[5] These binary structures could only be imposed and maintained through violence, hence the necessity of establishing military outposts on incorporated territories. The political, social, economic, and psychological repercussions of the colonial system have been criticized by many, but Fanon's searing critique was perhaps the most eloquent: "The colonial world is a world cut in two. The dividing line, the frontiers are shown by barracks and police stations." According to him, this inherently volatile and violent system could not be reformed, but only overthrown: "[C]olonialism is not a thinking machine, nor a body endowed with reasoning faculties. It is violence in its natural state, and it will only yield when confronted with greater violence."[6] As the colonial system "yielded" to one of sovereign nations, state structures were transformed both at the individual and aggregate levels. The binary divisions characteristic of the colonial era have given way to

---

[3] In the first volume of their trilogy, Michael Hardt and Antonio Negri contend that while the United States maintained a "privileged position," that "our postmodern Empire has no Rome." See Hardt & Negri 2000: 317.

[4] Tim Kane, "Global US Troop Deployment 1950–2005" 2006, The Heritage Foundation.

[5] The binary nature of colonial empires does not apply, of course, to all empires everywhere. For perhaps the most extensive discussion of the "politics of difference" within empires, see Burbank and Cooper 2010.

[6] Fanon 1963.

a more complex hierarchical structure in which national sovereignty is both bolstered and potentially undermined by multilateral military alliances.

Baseworld and the Protectariat

American defense planners, especially after 1945, have generally been obliged to negotiate any type of access or military presence with sovereign nation-states.[7] Whereas previous imperial powers often claimed ownership or stewardship over the lands and peoples they colonized, the United States renounced both. Even after Germany and Japan were defeated and occupied, Status of Forces Agreements (SOFAs), which regulated the exact nature of the U.S. presence, were signed. The creation of multilateral military alliances such as North Atlantic Treaty Organization (NATO) has also fundamentally transformed the nature of basing arrangements. Bases were henceforth not only created with the consent of the host country, but were also meant to serve common defense purposes. For both of these reasons, the American system was expected to be not only more legitimate, but also more stable than its British predecessor. Scholars wanting to emphasize the consensual or limited nature behind basing arrangements have referred to the U.S. basing system as an "empire by invitation" or a "leasehold empire."[8] Implicit in these depictions is that the U.S. military was "invited" at the behest of nation-states to settle in their countries, where they merely "lease" various facilities on a temporary, contractual basis. Built on such an apparently solid foundation, this force structure grew into a globe-encompassing network of military outposts, separate and largely isolated from much of the civilian sphere. As the Manichean colonial world was beginning to unravel, it was replaced by what appeared to be an equally binary division: the so-called Iron Curtain dividing East and West. International relations scholars were not incorrect to label the Cold War landscape as inherently bipolar, given the superpower confrontation with its uncompromising distinctions between friend and foe. But it was, in fact, more complicated than this. Especially on the Western half of this equation, a new logic of rule was being created.

NATO was established in Washington, DC in April 1949. Its raison d'etre was the provision of protection against a common enemy. In Article V of the NATO Treaty, the twelve founding members of the alliance affirmed that "an armed attack against one or more of them in Europe or North America shall be considered an attack against them all." Although the language of the NATO Treaty conveys the sense of equality and common interest between the twelve founding members, in reality the alliance has often been characterized by

---

[7] Many of the remaining British overseas territories are still used as military outposts today, including: Ascension Island, Bermuda, Diego Garcia, the Falkland Islands, and Gibraltar.

[8] Lundestad (2003) has referred to the United States as an "empire by invitation," whereas Sandars (2000) refers to the global deployment of U.S. forces as a "leasehold empire." For a critique of the notion that the United States was a "reluctant superpower" see Bacevich (2002).

hierarchy and divergent interests. NATO allies such as Turkey and Germany have been described as "security dependent" or as a "security consumer," as opposed to the United States, which is a "security producer."[9] As discussed in more detail later, it is important to underline that security is indeed "produced" through an international "production process" similar but not identical to global commodity chains. This notion of security is intentionally materialist as it emphasizes both the labor power and international linkages necessary in its production. In terms of the hierarchy embodied within NATO, perhaps most telling is that the head of NATO's military operations, who carries the title of Supreme Allied Commander Europe (SACEUR), has always been a U.S. citizen. Even the dynamic growth of many postwar European economies did not alter the fundamental situation of security dependence. Writing at the end of the Cold War, David Calleo described NATO as "essentially an American protectorate for Europe."[10] As the most powerful country in the Western alliance, and the bearer of nuclear weapons, it was the United States that was protecting Europe, and not vice versa. The NATO allies offered their land, airspace, infrastructure, and financial support, and in return they received American protection.

Rather than the binary divisions characteristic of the colonial empires, the American basing empire has instead created a structure that is tripartite both in a horizontal and vertical sense. The horizontal dimension consists solely of nation-states: the United States, the allies of the United States, and the declared adversaries of the United States. The vertical structure, however, includes societal actors as well: the U.S. military personnel stationed overseas, the host nation, and the so-called host population.

TRIPARTITE STRUCTURE – VERTICAL DIMENSION

TRIPARTITE STRUCTURE – HORIZONTAL DIMENSION

---

[9] For a critique of European "security dependence" on the United States, see, for example, Garnham 1994. For a more positive assessment, see Sherwood-Randall 2006.
[10] Calleo 1987: 3.

By guaranteeing the security of its NATO allies, the United States was taking on one of the key functions of the sovereign state. Using Weber's definition of the state as exercising a monopoly on the legitimate use of force within a given territory, NATO allies had clearly forfeited their "monopoly" of violence – or at least shared it – with the United States.[11] Of course, the U.S. government had no pretensions to represent the people in those countries that hosted American bases or to be responsible for their welfare – it merely claimed to protect them. Viewed from one perspective, this was an act of benevolence. From another perspective, it involved being subjected to a foreign military presence that operated outside the realm of the democratic polity.

When one state takes on one of the key functions of another state, interesting theoretical questions are raised in terms of both state sovereignty and state-society relations. The political science literature has examined how multilateral alliances such as NATO have affected state sovereignty, how and to what extent NATO may foster or undermine democratization processes, and how overseas military bases represent a Schmittian form of the *Ausnahmezustand*, or state of exception.[12] Sociologists and anthropologists have been more concerned with the human impact of the bases (discussed later). Although these scholars have produced rich empirical case studies, they have not adequately theorized what is perhaps most fundamental to this unique form of state-society relations. To approach this unexplored theoretical terrain, it is helpful to analyze the official discourse first.

The U.S. military refers to the non-U.S. citizens it claims to protect as the host population. This official terminology is problematic, however, as it implies a situation that is both temporary and based on consent, in contrast to the non-voluntary and perpetual nature of the U.S. presence in many parts of the world. The relationship between the U.S. military presence and the non-U.S. citizens under its security umbrella is inherently contradictory. Although the host population may be fully enfranchised citizens of their own government, they are at the same time disenfranchised by the U.S. presence. They have virtually no say in what the United States does on their territory, U.S. officials are not elected, and only rarely are U.S. personnel tried in local courts for any crimes they may commit. To capture the paradoxical nature of this relationship, I refer to the host population as base subjects or as a protectariat, as they are defined not by their relationship to the means of production, but instead by their relationship to the means of violence.

---

[11] Weber (1964) referred to this as the *Gewaltmonopol des Staates*.

[12] On how NATO has affected state sovereignty, see Milward 2001; on how alliances such as NATO can foster democratization, see Gibler and Sewell 2006; for a contrasting argument, see Reiter 2001. Giorgio Agamben (2005) has built on the German philosopher Carl Schnitt's discussion of the state of exception to analyze the post-9/11 world, with reference to the U.S. naval base at Guantanamo Bay. For a different reading of Schmitt and the Guantanamo Bay prison, see Johns 2005.

From Theories of Expansion to the Empirics of Contraction

A final comparative remark is in order before moving on to survey the existing literature. The growth of European overseas empires sparked one of the liveliest debates in the early twentieth century. Perhaps one of the most fought-over issues at the time among intellectuals, politicians, and revolutionaries alike was the question of imperial expansion. Hobson, Luxemburg, Lenin, Bukharin, and Kautsky all sought to understand why one European country after the other acquired colonies in Latin America, Africa, and Asia. Virtually all of the classic theories of imperialism grew out of this debate in the early twentieth century, which then inspired the next generation of scholars after World War II. In seeking to explain the nature of imperialist expansion, these authors put forth theories that ranged from emphasizing economic factors such as the search for resources in peripheral regions, the growth of monopolies, or the over-accumulation of capital to more political explanations such as those that focused on inter-imperialist rivalry or domestic strife in the metropole resulting from the growth of working-class militancy and socialist parties. Despite their political and intellectual differences, these early theorists had two things in common. First, they were all driven by a desire to explain imperial expansion, as opposed to the obverse process of contraction, which led them, secondly, to focus their analytical attention primarily on the metropole, and not the periphery.[13]

In a similar vein, historians in the United States have also debated the reasons behind U.S. expansion in the aftermath of the Spanish-American War in 1898, and have generally pointed to domestic political and economic factors.[14] Strangely, the much more dramatic expansion of the overseas basing network during the 1940s did not lead to a debate of similar intensity. It was presumed that bases were acquired during this period for a rather straightforward reason: to win World War II. Therefore, there was little need to develop intricate theories about American expansion during this period. And in any case, the United States had merely acquired bases, not colonies as after 1898. Accurate though this was, this led to the assumption that this collection of several thousand bases was merely a tool of U.S. foreign policy, or a means to an end. Even the most astute observers did not predict that this network of bases would take on a life of its own, creating a colossal military infrastructure as well as pockets of U.S. sovereignty scattered across the globe, while its raison d'etre can change at will according to directives from Washington. Maintaining the baseworld is no longer a means to an end, but has become an end in and of itself. In what follows,

---

[13] Luxemberg was the exception as the periphery featured more prominently in her theory. In *The Accumulation of Capital* she claimed that capitalism could only survive through the incorporation of non-capitalist regions into the world market. Luxemburg 1913.

[14] Historians such as William Appleman Williams, Walter LaFeber, Marilyn Young, and Gabriel Kolko are all referred to as "revisionists" because of their attempts to revise and replace the dominant discourse of American exceptionalism with evidence of, and explanations for, American expansionism.

I do not put forth a theory regarding the expansion of the American baseworld; instead, I intend to empirically analyze its limits. As is documented in the following chapters, these limits were rarely if ever set by economic constraints or Congressional oversight or any other dynamics originating in the United States, but rather by social unrest in the countries that host U.S. military bases.[15]

Understanding the limits of the American empire requires that we first establish a degree of conceptual clarity before moving on to summon the empirical evidence. First, we need a clear understanding of both the continuities and novelties of the American overseas basing network when compared with previous imperial formations. Instead of colonial sovereignty, a new form of layered or double sovereignty came into existence as the U.S. military presence existed alongside the formal sovereignty of the nation-states that hosted it. Instead of the binary divisions of colonial empires, the international system evolved into a tripartite hierarchy along both horizontal and vertical dimensions, both during and after the bipolar confrontation of the Cold War. Instead of colonial subjects, the host populations were fully enfranchised citizens of their own government, but at the same time disenfranchised vis-à-vis the U.S. presence. To capture the contradictions inherent in this situation, I refer to them as base subjects or as a protectariat as they are defined not by their relationship to the means of production and in no sense an economic class, but rather by their relationship to the means of violence. Base subjects, like citizens, are therefore better understood as a Weberian status group. It is their agency that would be critical in determining the limits of the baseworld. Therefore, we must place the emergence and evolution of opposition to the American military presence at the center of our analysis.

Despite the vastness of this overseas network, little is known about it. Although the American military presence in war zones may receive media coverage and be of some general interest, the American public is often unaware of the fact that the global U.S. military presence is neither temporary nor confined to combat zones. The incognizance of many Americans contrasts sharply with the perception in host countries, where basing issues can often overshadow other aspects of the bilateral relationships. For many U.S. allies, American troops and bases are the very symbol of their country's relationship with the United States.

State of the Field – Existing Literature

The small but growing literature on overseas bases can be divided into macro-level studies within an international relations framework and micro-level case studies of individual host countries that are more sociological or

---

[15] Beginning in the 1970s, a literature emerged on the "decline of American hegemony." Much of this focused on either the defeat in Vietnam or the rise of Japan and the European Union as economic competitors. In other words, the U.S. decline was analyzed as being primarily the result of losses in the military or economic sphere, but never in terms of losing access to territory overseas. To emphasize the importance of territory to the American empire, rather than discussing the "rise and fall" of the American baseworld, I prefer instead to refer to its geographical expansion and contraction.

anthropological in nature. These two bodies of literature tend to ask different questions and be directed toward different audiences. Whereas the international relations literature is interested in how states optimize the projection of force and how overseas bases can be made more effective, the anthropological or sociological literature often tends to be more interested in the human impact of the bases and asks how bases can cause less harm. Because a base's utility depends partly on its interoperability with other facilities, the international relations literature generally views individual bases as part of the larger network of American installations. The literature that investigates the human interactions on and around bases, on the other hand, looks more closely at the local situations, while often ignoring the larger network of which any particular base is a part. Until now, there has been little dialogue between the two literatures.

Although U.S. facilities have been spread far across the globe, research has been done on only a handful of countries that host U.S. bases. Thus far, the literature has focused primarily on non-European countries and territories such as Puerto Rico, Cuba, Diego Garcia, South Korea, New Zealand, Japan/Okinawa, and the Philippines. Europe, however, has in fact hosted more U.S. troops than any other region of the world. During the Cold War, the United States established military bases in fourteen separate European countries, including West Germany, Britain, Belgium, the Netherlands, Luxemburg, France, Italy, Portugal, Spain, Greece, Turkey, Denmark/Greenland, Norway, and Iceland. Considering that since 1950, 41 percent of all U.S. troops were deployed in Asia and 52 percent were deployed in Europe,[16] the fact that European host countries have been understudied is curious indeed.

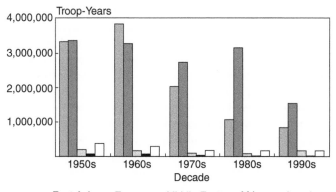

Cumulative U.S. Troop Deployments by Region and Decade

□ East Asia ■ Europe □ Middle East ■ Africa □ Americas

---

[16] Kane 2004.

Although the socio-anthropological literature often assumes a more critical perspective, it still tends to maintain a residue of U.S.-centrism in that it asks what U.S. personnel do or how U.S. policies affect an area; in other words, the U.S. presence is seen as possessing a primary agency regardless of how this impact is assessed. For this reason, despite the diversity of perspectives and opinions, many scholars have not taken seriously the possibility that local conditions or agencies could impact the network of U.S. bases and its adherent policies and strategies. Chalmers Johnson came closest when he appropriated the CIA-term "blowback" to warn about the possible consequences of U.S. foreign policies. However, even he did not offer a systematic explanation for why anti-base movements emerge. Neither do the existing case studies that include rich narratives of the local situation and discussions of why local people accept, acquiesce, or challenge the foreign military presence. Furthermore, although they have shown that a foreign base can have a negative impact on the host population, few have demonstrated that problems caused by bases can, in turn, impact U.S. basing strategy, in particular the ability of the United States to locate bases wherever it desires.

Based on the existing literature, it is possible, however, to make a few very broad generalizations about why resentment to a foreign presence may emerge. In the literature on Puerto Rico, Cuba, and the Philippines, all of which were acquired as a result of the Spanish-American War in 1898, the neocolonial relationship between the United States and the host country (or territory) is often highlighted as explaining local resentment toward the American presence.[17] This was perhaps most evident in the Philippines, where the U.S. government agreed to grant independence to its largest colony in 1946, under the condition that, in return, it receive ninety-nine-year rent-free leases to the Subic and Clark bases along with fourteen other facilities. The United States maintained sovereignty over the bases and even the city of Olongapo, and jurisdiction over all Filipinos employed on the bases.[18] In Japan, Okinawa, and South Korea, the widespread sex industry that has grown up around U.S. military installations is often cited as one of the main factors that sours bilateral relations with these countries. Military prostitution, the development of sex tourism, children of American servicemen who grow up fatherless, and violence toward women can all be traced to the U.S. presence.[19]

Whereas in many parts of the world the U.S. presence was bilateral, with the United States being the only foreign military presence, the American presence in Europe was embedded in a multilateral framework, with many European countries hosting small contingents of NATO personnel from a number of different places. A brief comparison of the postwar American presence in

[17] Simbulan 1985; Ricardo 1994; McCaffrey 2002; Barreto 2002; Baker 2004.
[18] Simbulan 1986.
[19] Stoltzfus and Sturdevant 1992; Moon 1997; Cornwell and Wells 1999; Johnson 1999; Enloe 2000; Höhn and Moon 2010.

Japan and West Germany is illustrative. In Japan, the United States was the sole occupying power after World War II, and Okinawa was a protectorate of the Pentagon until it reverted to the Japanese mainland in 1972. In West Germany, by contrast, the American commanders shared authority with their British and French colleagues. Therefore, the multilateral framework of NATO was assumed to provide checks on the exercise of power.[20]

If the American system of signing mutually beneficial basing agreements with sovereign nations was more stable than the British system established with colonial dependencies, then the American presence in Europe should have been the most stable of all, when compared to other regions of the world. It was not burdened with a colonial past, and was built on the premise of common defense within NATO. From the American perspective, a deployment to Europe usually meant the opportunity to bring spouses and children rather than endure long separations from family members. Finally, U.S. military personnel could enjoy a standard of living similar to or even higher than what they were accustomed to within the United States. For this reason, Europe, and Germany in particular, became the favorite overseas post for generations of U.S. military personnel. From the European perspective, the American presence offered a welcome guarantee that the conflicts that had engulfed Europe in the first half of the twentieth century would not be repeated, while at the same time preventing any possible Soviet encroachment. The U.S. presence in Germany in particular served a dual purpose: it offered protection to the Germans against the Soviet Union while offering protection to all other NATO allies against the potential reemergence of German militarism. In short, the U.S. presence was initially regarded as a public good of which virtually no one in the Western alliance disapproved.

However, opposition to the U.S. military presence arose both when and where it was least expected: in Europe during the height of the Cold War. This happened even though it was assumed that fear of the Soviet Union and membership in the Atlantic Alliance would assure loyalty to the United States and its security umbrella. This opposition was not confined to marginal groups but became widespread and took a number of different forms. The rise of oppositional movements against the U.S. presence in Europe is therefore perhaps the most difficult to explain. At different times during the postwar period and for different reasons, France, Denmark, Iceland, and Yugoslavia requested that the United States either withdraw its military presence entirely, or severely circumscribe its activities. Short of ousting the American presence, other countries denied the United States access to the bases on its territory or imposed other types of access or overflight restrictions at various critical junctures during the Cold War. Spain, Portugal, Greece, and Turkey began demanding payment for basing access that had previously been freely given.[21] Even wealthy countries such as Germany began to refuse to subsidize the U.S. presence as it had

---

[20] For a comparison of the role of Germany and Japan in the "American imperium" including a discussion of multilateralism in Europe and bilateral relations in Asia, see Katzenstein2005.
[21] McDonald and Dendahmane 1990.

done before.[22] Finally, social movements in a number of countries including Great Britain, West Germany, the Netherlands, Italy, Turkey, and Greece slowly inhibited the ability of the U.S. military to operate and even carry out normal practice maneuvers. Although rising costs can be calculated and the impact of overflight restrictions is a straightforward matter, assessing the impact of social unrest is more difficult. This is especially true given that the anti-base opposition did not only take the form of social movements, but included a wide range of contentious politics, as discussed shortly. Yet the general trend was clear: over the course of the Cold War, the number of overseas bases had declined and the tangible and intangible costs of maintaining those that remained had risen.

By 1990, the accumulation of these problems led James Blaker, a former defense advisor, to publish a book titled *United States Overseas Basing: An Anatomy of the Dilemma*. He argued that the primary dilemma of the overseas force structure was that "further reductions of overseas bases, or additional constraints on what can be done from them, threaten the national security interests of the United States."[23] Other observers in the late 1980s and early 1990s described similar trends. Robert E. Harkavy referred to a "decoupling of basing ties" and John Woodliffe described different factors that "cumulatively point to a marked contraction of the existing network of foreign military bases."[24]

Although international relations scholars like Blaker and Harkavy should be credited with having detected the overall trend of increased restrictions and its implications for U.S. policies in the late 1980s, they have not done much in the way of explaining the "decoupling of basing ties" or the social forces that gave rise to this phenomenon. This can most likely be attributed to two reasons. First, NATO membership is assumed to guarantee a certain amount of cooperation. If for some reason this cooperation is withdrawn, it is assumed that this is the result of changes in the domestic political situation of the host, which is seen as an independent variable, over which the United States has no control. Secondly, from a strictly American perspective, this phenomenon was in some ways inexplicable because many European allies were seen as being the primary beneficiaries of the security umbrella while the Americans were shouldering the burden. As Catherine Lutz has pointed out, the official language about U.S. overseas bases tries to present them as "gifts" to host nations.[25] During the Cold War, journals such as *Foreign Affairs* featured articles that explained the rise of anti-Americanism in Europe as a type of ingratitude, which resulted from the situation of security dependence.[26] More

---

[22] Zimmermann 2002.
[23] Blaker 1990.
[24] Harkavy 1989; Woodliffe 1992.
[25] Lutz 2006.
[26] See, for example, Howard 1982/1983.

than twenty-five years later, it is still possible to find echoes of this argument, as when Robert Kagan suggests that Europeans are simply not in the business of exerting hard power, and therefore neither appreciate it nor understand it.[27] Simon Duke's encyclopedic *United States Military Forces and Installations in Europe* contains valuable information on all fourteen European host nations and acknowledges that European allies are not just benefiting from the security umbrella, but also shouldering part of the burden. He does not, however, offer any explanation for challenges to the U.S. presence.[28] Since the early 1990s, two edited volumes have been published that include contributions on anti-base movements, but they lack systematic comparisons because they are collections of essays written by various authors.[29] Chalmers Johnson's trilogy examines the changing nature of U.S. power, in which overseas bases figure prominently. The first book of the trilogy addresses U.S. military power in Asia and predicts that the United States would suffer unintended consequences or "blowback" from its misguided policies, as mentioned briefly earlier. The second and third books of the trilogy expand this argument beyond the focus on Asia to show how U.S. bases in particular provoke varying forms of resentment in different parts of the world. His discussion, however, was not geared toward producing testable hypotheses about why anti-base sentiment does or does not emerge.[30]

By comparing a number of host nations, thus far, Kent Calder, Alex Cooley, and Andrew Yeo have made the most systematic attempts to understand the circumstances under which basing agreements are likely to be contested. Cooley argues that the regime of a host country is most likely to challenge the U.S. presence when it is undergoing a phase of democratic transition, especially if the basing agreements were made prior to the transition to democracy.[31] In other words, he tends to assume that criticism of the U.S. presence emerges because of the politics of democratic transitions, which may be entirely unrelated to the way in which the U.S. military operates. Calder agrees that regime-shift in the host country often leads to trouble for the U.S. presence, but he also identifies two other sources of tension, namely base-community contacts (what he calls the Contact Hypothesis) and a previous colonial experience (the Colonization Hypothesis). His other two hypotheses are that the United States often supports dictators in nations hosting bases "to a degree greater than in American foreign policy more generally" (the Dictatorship Hypothesis) and that a liberating occupation generally leads to a stable basing environment (the Occupation Hypothesis).[32] Both Calder and Cooley have substantially improved our understanding of the politics of overseas bases. Both have identified regime-shift in the host country as the primary cause for changes in basing

---

[27] Kagan 2004.
[28] Duke 1989.
[29] Gerson and Birchard 1991; Lutz 2009.
[30] This trilogy includes: *Blowback* (2000), *Sorrow of Empire* (2004), and *Nemesis* (2006).
[31] Cooley 2008.
[32] Calder 2007: 225ff.

patterns, whereas technological developments, changes in strategy, and realignments in the international landscape are less likely to lead to withdrawal or other major shifts. Considering the spread of democratization since the end of the Cold War and the expansion of the NATO alliance, it would seem that their theories can only be applied to a shrinking number of cases. Assuming that radical regime change is unlikely to happen in democracies, what other factors could influence basing decisions? For our purposes, their theories, although helpful in other contexts, cannot fully explain the existence of social movements from below in democratic regimes, nor can they explain changes in the nature of U.S. power over time.

Andrew Yeo has argued that it is the level of elite security consensus that is most decisive in determining the outcome of anti-base movements. He posits that a strong security consensus among the elites of a host country will make it difficult for anti-base activists to achieve their goals, whereas a weak security consensus among elites allows activists to potentially influence policy outcomes. Through an impressive array of case studies of anti-base movements in the Philippines, Japan, Italy, Ecuador, and South Korea, he argues that the variation in movement outcomes was primarily determined by the level of elite security consensus within the host nation. His analysis is based on the assumption that anti-base activists can only hope to achieve their goals through influencing or convincing their elected officials. Although it is true that the host government often acts as an intermediary between the local population and the U.S. military presence, and that some activists appeal to their representatives in an attempt to convince them to make certain changes in the basing agreements, not all activists choose this route. In my case studies of Germany and Turkey, I have found that certain protest activities had a *direct* impact on the U.S. military presence, and essentially circumvented host-nation elites. Other protest activities were catalysts for changing the level of elite consensus within the host government. And at times, U.S. military officials were forced to respond to certain types of unrest or accept restrictions on their ability to operate, although these changes did not necessarily entail official revisions to the Status of Forces Agreements and therefore did not involve negotiations with host-nation elites. I have also employed a broader notion of contentious politics than the scholars discussed previously, and have included in my analysis any significant form of social unrest that took issue with the U.S. military presence, including parliamentary opposition, civil disobedience, labor disputes of base workers, and violent attacks against U.S. bases and personnel by groups engaged in armed struggle. Rather than narrowing my focus to anti-base movements per se, I have attempted to include any form of social unrest that represented a challenge to the U.S. military presence in Turkey and Germany from 1945–2005.

Although generally from a more limited geographical perspective, scholars of social movements have made significant contributions to understanding the impact of social unrest on political processes. Until now, social movement scholars have generally compared movements with similar agendas that developed in similar contexts. For example, the German peace movement in the 1980s is often compared to the Campaign for Nuclear Disarmament in Great Britain or the peace movement in the Netherlands because they all developed in opposition to the deployment of a new generation of missiles.[33] Fewer attempts have been made to compare movements that developed in Northern Europe with those that emerged in the Southern tier.[34] Finally, these movements have more often been analyzed in terms of their impact on domestic political processes, rather than in terms of their relationship to the American military presence, although it was often one of their primary targets.

THE PUZZLE

My research will address these shortcomings in the literature by examining the causes and consequences of social unrest against U.S. military bases. I will seek to answer several questions:

- Within NATO, U.S. military bases were founded on the principles of national sovereignty and common defense goals and therefore widely regarded as not only legitimate, but even as a necessary form of protection vis-à-vis the Soviet Union. And yet anti-base movements emerged in allied European countries during the height of the Cold War. Why? What were the causes of opposition to the U.S. presence?
- What form did this opposition take?
- What were the consequences of the anti-base unrest? How did the U.S. military respond to these challenges? To what extent was the U.S. presence in Europe affected by the rise of oppositional movements?
- In Chapter 4, I examine the post-Cold War period. After the collapse of the Soviet Union, the ostensible threat that necessitated a U.S. presence in Europe for more than half a century no longer exists, and yet the military bases remain. During this period, the most vigorous opposition to the US presence emerged during the run-up to the US-led invasion of Iraq. Despite similar levels of mobilization, basing access was granted in Germany, but denied in Turkey. Why?

EXPLANATORY FRAMEWORK – CONCEPTS FOR ANALYSIS

Understanding the Causes of Anti-Base Unrest

To understand opposition to the presence of an allied and allegedly friendly military, I draw from several sources. The U.S. basing network was established during the waning hours of the European empires. Weakened already by the devastation of World War II, the foundations of British, French, Dutch, Belgian,

---

[33] Johnstone 1984; Cartwright and Critchley 1985; Kitschelt 1986; Joffe 1987; von Bredow 1987.
[34] Some exceptions are Wittner 2003 and Della Porta 1995.

and Portuguese colonialism were further undermined by growing nationalist and independence movements in the Third World. The United States could claim legitimacy precisely because it renounced the forms of ownership that formal European colonialism entailed. Although many scholars have compared various types of imperial formations which allow for greater detail and complexity, here I follow Engseng Ho's schematic categorization of three phases of empire: (1) the earliest settler colonies in the Americas in which both land and labor belong to the empire as colonies and slaves respectively, (2) colonized metropoles in Asia and Africa in which the land but not the labor belongs to the empire, and (3) anti-colonial empires in which neither land nor labor belongs to the empire. "Just as critiques of the first European empires underwrote the second, critiques of the second empires, for being colonial, underwrote the third. The U.S. empire supplanted its predecessors at the end of World War II, when it pushed a devastated Europe to decolonize and supported independence movements against them."[35] The U.S. basing network belongs to the third category of anti-colonial empires and takes the specific form of an empire of overseas bases. In the postcolonial world of bases, the territories are sovereign and the people are free. The legitimacy of this system derived not only because it renounced previous forms of colonialism, but also because it could claim to offer protection against an external threat. Indeed, the two issues were intertwined. One of the reasons the United States was at times reluctant to support the continuation of colonial rule was because this entailed diverting military resources out of Europe. Regarding discussions on NATO in 1949, Secretary of State Marshall said: "[W]hen we reached the problem of increasing the security of Europe, I found all the French troops of any quality were all out in Indochina, and I found the Dutch troops of any quality were out in Indonesia, and the only place they were not was in Western Europe."[36] In short, Washington wanted its European allies to be more focused on Europe. The creation of NATO involved redrawing the map of Europe, with new boundaries between "insiders" and "outsiders." Previous enemies (the Axis powers of Germany, Austria, and Italy) and those who remained neutral for most of the war (Turkey until February 1945) were included as insiders, whereas former allies (the Soviet Union) were declared outsiders. Those within NATO were to be protected; those outside were to be threatened.

As the map was redrawn, ideologies were reinvented. If previous ideas ranging from blatantly racist discourses of cultural superiority or white supremacy to the seemingly more benign notion that European civilization was more "advanced" than other parts of the world (which had underpinned much of European colonialism) had not been entirely rooted out, they had at least fallen somewhat into disrepute. Or had they? At least for the duration of the Cold War, the old notion of Western supremacy lived on in the new ideology of anti-

---

[35] Ho 2004: 227.
[36] Quoted in Kolko 1972: 558.

communism. Emboldened with a newfound and seemingly revitalized sense of Western superiority, the United States and its European allies convinced themselves that they could (if necessary) inflict harm on – or even annihilate – the peoples of Eurasia. Often disguised under the rhetoric of defense, the idea of harm needs closer scrutiny.

Andrew Linklater has distinguished between two types of transnational harm: concrete harm, which is harm that actors intend to do to designated others, and abstract harm, which spreads across borders unintentionally.[37] Warfare is perhaps the quintessential form of concrete harm. The problem, however, was how to demonstrate that one could potentially inflict harm on the USSR without invading or violating the borders of the Iron Curtain. The solution to this was to amass a huge military presence as close as possible to the Warsaw Pact countries, but still within the geographical boundaries of NATO. Germany and Turkey were both frontline states and therefore central to this endeavor. The same military bases that were to execute harm to outsiders were to represent protection to insiders. Official Defense Department publications underline the fundamentally contradictory nature of the overseas military presence, such as this statement in a publication of the Navy/Marine Corps: "We will assure our friends and allies, and together with the U.S. Air Force, U.S. Army, and U.S. Coast Guard, we will dissuade, deter, and defeat our nation's enemies."[38]

The theoretical framework that Charles Tilly developed to analyze European state formation can help us understand the contradictory nature of protection. According to Tilly, at their core, states are in the business of providing protection from both internal and external threats, and hence state-making and war-making are inseparable historical processes. He distinguishes between legitimate protection (a ruler who provides a shield against a truly dangerous adversary) and a protection racket (a ruler who provides a shield against a threat that is either imaginary or that arose as a result of his own activities). He describes legitimate protection as a "comforting" form of protection, and he describes a protection racket as "ominous." In other words, whether or not the protection is legitimate depends to a great extent on the source and nature of the threat.[39] Following Stinchcombe, Tilly adds that legitimacy also depends on the consent of other power-holders. This stands in contrast to those following the Rousseauian argument that a state gains legitimacy through the consent of the governed. Either way, the fundamental point is that legitimacy is relational and not something that can be unilaterally declared. For the current project, the unit of analysis that is most relevant is that of the host country. Because citizens of host countries are not "governed" by the U.S. military

[37] Linklater 2002: 171.
[38] This document was coauthored by Secretary of the Navy Gordon England, Chief of Naval Operations Vern Clark, and Commandant of the Marine Corps James L. Jones: "Naval Power 21 ... A Naval Vision" 2002.
[39] Tilly 1985.

presence, the Rousseauian notion of a social contract does not apply. And yet the issue of legitimacy is still relevant. Rather than legitimacy depending on the "consent of the governed," for our purposes we could say that it depends on the "consent of the protected." Tilly emphasizes that consent is more likely to be given if a state can live up to its claim of offering protection. "A tendency to monopolize the means of violence makes a government's claim to provide protection, in either the comforting or the ominous sense of the word, more credible and more difficult to resist." [40]

Arrighi applies this analogy to the field of international relations and argues that during the first two decades of the Cold War, the United States offered legitimate protection to its allies because it provided inexpensive protection against a perceived danger that the United States had not produced itself, namely the USSR.[41] However, during the Vietnam War and with the ensuing balance of payments deficit, the United States began demanding various forms of tribute from its allies. Within NATO, this took the form of burden sharing as allies were asked to share the financial burden of U.S. protection. In the face of U.S. defeat in Vietnam, U.S. protection henceforth not only came with a price tag, but was also less credible. After a decade of détente, U.S. protection also seemed less necessary to many European allies. For Arrighi, therefore, the gradual withering away of legitimate protection was part of both the larger crisis of American hegemony as well as changing relations between the superpowers, but cannot be fully explained by these events alone. Although Arrighi and others are certainly correct in pointing to Vietnam as a turning point in many ways, my analysis shows that opposition in Turkey and Germany was most fierce regarding how the U.S. military operated *in those countries*, not in Vietnam. We are therefore dealing with a distinctive form of social unrest, best understood as anti-base contention. For the United States, the war in Vietnam provoked a major anti-war movement that in many ways changed the fabric of American society. But U.S. allies in Europe were wracked by *both* anti-war and anti-base movements. For those societies and for the U.S. troops stationed there, the anti-base movements may have in fact been the more significant of the two. And yet, because this type of social unrest does not fit neatly into commonly held understandings of social movements, it has not been fully understood or sufficiently analyzed. For defense planners, the lessons from the Vietnam War and the unrest that it unleashed were confined to two main issues. First, after being defeated by Vietnamese peasants, the U.S. military desperately needed to regain its credibility as a military superpower. Second, it was deemed necessary wherever possible to shield itself from a casualty-averse public. This realization led to a transformation of the way the United States conducts warfare, leading to a professional military that preferred to engage in limited combat operations or covert actions, and an increased reliance on unmanned aircraft such as

[40] Ibid: 172.
[41] Arrighi 2005; see also Arrighi 2007: Part III.

drones. It did not lead to a transformation of the overseas military presence.[42] It was opposition to the American military presence in allied countries, not the Vietnam War, that eventually led to changes being made. However, for a variety of reasons, perhaps because it took various forms rather than being a single unified movement, or perhaps also because of the general lack of analysis, the anti-base unrest did not result in straightforward "lessons" being drawn, as in the case of the Vietnam War. Prior to the Vietnam War, American male citizens were subject to the draft; one result of the anti-war movement was to abolish the draft so that U.S. citizens could not be forced to fight against their will. In a similar vein, citizens in countries that host U.S. bases – the protectariat – may feel that they have been "drafted" into being subjects of the protection regime against their will, as they have neither voice opportunities through official channels nor exit options, short of emigration.

In her analysis of the transformation of U.S. hegemony, Silver adds the dimension of popular support, which is key for my own research. She also follows the line of argument staked out by Tilly and underlines that the process of war-making and state-making were intertwined for another reason as well: as warfare became industrialized, larger segments of the population became "critical cogs in war machines" and it thus became necessary for rulers to ensure popular support for the war effort. The extension of the franchise and concessions to labor unions can be understood in this context of tying the masses to the nation-state to secure their loyalty. This process, in turn, increased not only the nationalist sentiment, but also the bargaining power of workers.[43] In other words, popular resistance to states was important because it obliged rulers to grant certain rights to their citizen-soldiers, which in turn later constrained the ability of states to make war. Although these constraints certainly may have affected the foreign policy of states, Tilly's and Silver's analyses are focused on the domestic arena of the nation-state and the contested relationship between governments and their national armies. To understand the complex dynamics between a foreign military presence and the population that "hosts" it, we must leave this terrain.

The Production of Violence and Protection

When a state shifts its war-making capabilities (i.e., military bases) onto foreign territory, it is no longer necessary to grant concessions to noncitizens, hence the imperial state appears to be less constrained in its actions. This, of course, does not mean that the problems associated with war-making have been solved, but merely that they have been shifted onto the client states that host U.S. forces. In fact, this was one of the reasons it was believed that the American empire

---

[42] Of course some of those facilities that were no longer needed for combat operations in Vietnam were shut down, but these changes did not entail a rethinking of the way the overseas U.S. presence operates in allied countries.

[43] Silver 2004: 19–38.

would be more stable than its European predecessors. Precisely because it left the task of governing to the host country and was devoid of a civilizing mission or any pretentions to represent or be responsible for the welfare of the people it claimed to protect, it had rendered itself outside the democratic polity of its host countries. It was assumed, therefore, that it would not be vulnerable to pressures from below. By renouncing responsibility for political representation and economic redistribution, perhaps the two single most common demands of social movements both in the periphery and in the metropole, the United States had ingeniously insulated itself from the contentious politics that had defined the modern era. The United States was able to project power far beyond the homeland, and therefore had all the benefits of an imperial presence without any of the burdens of a colonial empire. Or so it seemed. As we will see, the very strength of the U.S. baseworld would eventually reveal itself as its greatest weakness.

By attempting to insulate itself from popular pressure the United States had in a sense "disenfranchised" the very people it claimed to protect. In this sense, citizens of sovereign nations that host a U.S. military presence face a similar dilemma to the subjects of non-sovereign colonies: how to voice their concerns regarding the U.S. presence in their midst as it exists outside the realm of their democratic polity. Equally difficult is the issue of how to pressure their government to abrogate or even modify the existing basing agreement if it is obliged to honor the agreement as a NATO member. In fact, as discussed previously, Cooley has argued that consolidated democracies are *more* likely to honor basing agreements than regimes that are in the process of transitioning to democracy, which means they are *less* likely to respond to pressures to modify or terminate such agreements. Because voice opportunities through official channels are lacking, social movements arise in their place. Precisely because the political opportunity structure was closed, grievances could only be expressed outside of institutionalized settings. The issue of grievances and the emergence of anti-base movements is discussed in greater depth in the next section. For the moment, I hope to have outlined the vertical power relationships between the U.S. military presence and the non-American citizens it claims to protect in allied countries. We must also, however, consider the horizontal power relations of the U.S. basing empire.

In its totality, the overseas network of U.S. military bases both threatens and protects at the same time. I argue that this network forms a global division of labor that produces both violence and protection. Just as production processes occur through an international division of labor, with commodity chains linking factories, warehouses, and retail stores around the world, so too is protection and violence "produced" through the global deployment of U.S. troops and bases. The production of violence and protection, however, requires more than just military facilities. At a bare minimum, landing a plane at an airbase requires access to airspace. Moving supplies or personnel from one facility to another requires a network of roads. Naval vessels require ports and harbors, both for ship repair and for rest and recreation. In some cases, particularly when military personnel are accompanied by their families, or live off-base, the level of dependence on the local economy and civilian infrastructure may increase substantially. But even when military personnel are secluded to their bases and do not intermingle with the local community, the facilities themselves are never entirely self-sufficient but rely on the infrastructure of the host nation to function. In short, bases are never just bases, but must be understood as a military *presence* that expands beyond the perimeters of any given facility. To capture the political and sociological nature of the military presence, I use the term "protection regime."

Finally, this global division of labor takes the form of different types of protection regimes within different countries that host U.S. bases. Just as factory regimes in different locations take on different forms, so too do protection regimes vary. Michael Burawoy has categorized some factory regimes as "hegemonic" and others as "despotic." I argue that some protection regimes are hegemonic (i.e., they represent legitimate protection), whereas others operate as a protection racket and are fundamentally coercive in nature.[44] Just as the same multinational corporation may operate plants in different parts of the world, which may be characterized by vastly different working conditions or different developmental impacts on the local community, so too is it possible that the same military force may operate bases in different parts of the world whose impacts may vary greatly from one place to another. In other words, whereas it may be possible to make generalizations about U.S. hegemony at the global level (i.e., that it is either rising or declining), it is not possible to make such generalizations about the protection regimes, because they vary across both time and space.

To understand anti-base movements, it is necessary to move away from static and a priori conceptions of U.S. power. My over-arching hypothesis is that the U.S. basing system can take on various forms, including what I have termed precautionary protection, legitimate protection, pernicious protection, and a protection racket. Although the force structure offers protection to U.S. allies, it maintains the threat of violence toward others and is therefore Janus-like: both benign and belligerent, both creating and undermining security. Contradictions are inherent and built in to the system of overseas bases. It is also dynamic and can change over time. For this reason, even in those countries that it claims to protect, public opinion toward the U.S. presence can be ambivalent.

I will build on Tilly's conceptualization of war-making and state-making, by arguing that for a foreign military presence to qualify as offering legitimate protection toward a host nation, it must fulfill two minimum requirements: (1) it must offer protection against an outside threat which it did not create itself, and (2) it must cause no harm toward those it claims to protect. If both of these conditions are met, I argue that the foreign military presence constitutes

[44] Burawoy 1983.

legitimate protection. This means, of course, that the loss of an external threat constitutes a loss of legitimacy for the foreign presence, through no fault of its own. Although this conceptualization could potentially be applied to any military stationed on foreign soil, this study will be focused on the U.S. presence overseas.

In other words, whereas for Tilly the primary independent variable was the nature of the outside threat, in my conceptualization, there are two independent variables: (1) the level of outside threat, and (2) the level of collateral harm caused by the U.S. presence. The dependent variable in this study is the emergence of unrest against the U.S. military presence. See Figure 1.3 at the end of this chapter, which demonstrates that the level of outside threat can be classified as high or low and the level of collateral harm caused by the protector can also be categorized as high or low. The narrative of course contains more details and nuances, but Figure 1.3 illustrates the broad strokes of the historical dynamic. During the Cold War, for example, the outside threat against which the United States claimed to be protecting was the Soviet Union. Although the Cold War was characterized by tensions which waxed and waned over time – as there were both phases of détente and phases of escalating tensions, sometimes climaxing in major crises such as the Cuban Missile Crisis in 1960, in my model I assume that the level of threat during the Cold War remained constant as long as the superpower confrontation dominated the political landscape. This means that for the U.S. presence to qualify as legitimate protection, one of the two criteria was already fulfilled by the existence of an outside threat.

Some readers may contend that the Soviet threat was exaggerated for political reasons, or even that the USSR was never truly a threat in the first place, and that its foreign policy vis-à-vis NATO was more reactive and defensive than aggressive. Whether or not the Red Army ever seriously plotted an invasion of Western Europe is a question that can only be definitively answered through archival research in Moscow. It is not my intention, however, to uncover Soviet motivations. Nor am I making any judgments about the quality of the Soviet army, air force, or navy. If historians were ever to conclusively agree that the Soviet Union never intended to encroach upon West Germany, Turkey, or any other NATO country, this would not change the fact that *during the Cold War* defense planners operated under the assumption that the USSR represented a threat – either politically or militarily or both – and that it was necessary to plan accordingly. In my model, therefore, I assume that there was a high level of external threat during the Cold War. With the dissolution of the Soviet Union, the external threat disappeared, and therefore the level of external threat to Germany and Turkey is characterized as low.

Moving back in time for a moment, it is important to consider the few years between the end of World War II in 1945 and the emergence of the Cold War in 1947/1948. During this time, the external threat to Europe was relatively low, as the Wehrmacht had already been defeated and the Soviet Union had not yet emerged as a threat. For this reason, the large wartime military presence was demobilized and many troops were sent back to the United States. However, approximately 70,000 U.S. soldiers remained in Germany as part of the occupation. I have termed this period a "precautionary protection regime" as the primary purpose of the military presence was to guard against any possible reemergence of German aggression. The U.S. soldiers, together with their British and French counterparts, remained in West Germany more as a precautionary measure than because of any immediate external threat. This is discussed in greater detail in Chapter 3. The combination of a low level of external threat and a low level of collateral harm can therefore be considered a "precautionary protection regime" and is in the lower left quadrant of Figure 1.1. This does not mean, however, that all military occupations would fall into this category. In theory, they could fall into any of the four categories of protection regime, depending on the level of external threat and collateral harm.

Collateral Harm

As described previously, the second criterion depends on the U.S. presence causing no harm toward those it claims to protect. It is of course possible that U.S. troops engage in other activities such as humanitarian aid and thereby also acquire or even bolster their credibility. However, I utilize a minimalist definition that only requires that only the two basic conditions defined earlier are fulfilled. Thus, when both conditions are fulfilled, I classify the position of the U.S. military as representing legitimate protection. In this sense, it can be considered hegemonic, defined in the Gramscian sense as being a combination of coercion and consent. In other words, it enjoys some degree of active or passive consent among its protégés, or at least suffers no organized resistance.[45]

At times, however, the legitimacy of the U.S. presence may be called into question. The U.S. presence may begin to appear threatening, perhaps even more so than the threat against which it claims to protect. This might happen because it undermines national sovereignty, because the stationing of nuclear weapons arsenals poses a threat to local citizens, or because the level of environmental damage becomes intolerable. At other times the behavior of U.S. troops, including racial or religious discrimination or sexual harassment, may cause offense or even pose a threat to the citizens of the host country. I refer to these grievances as representing "collateral harm" as they neither fall into the categories of "abstract harm" or "concrete harm" as outlined by Linklater, and which were discussed previously.

I argue that, even though the threat of a potential Soviet attack on Europe still existed as long as the Cold War lasted, the threat posed by the U.S. military presence with its seemingly permanent foothold within Europe, began to appear both imminent and real. The presence of large numbers of U.S. military personnel who operated with considerable autonomy as they could be neither

---

[45] According to Gramsci, hegemony is "characterized by the combination of force and consent, in variable equilibrium, without force predominating too much over consent." See Gramsci 1991.

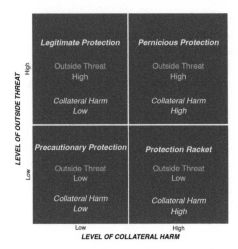

FIGURE 1.1. Generic protection graph.

elected nor held accountable for crimes in local courts, began to seem like a threat in itself. I have termed this type of protection regime "pernicious protection." Although it is technically possible for opposition to emerge during any of the four different types of protection regime, it is after the transformation to pernicious protection – or when collateral harm occurs – that consent for the U.S. presence is most likely to erode and anti-base opposition emerges as a result.

If the external threat ceases to exist and the yet the U.S. presence continues to exist and even continues to cause collateral harm to its protégé, it can be considered a protection racket. This situation is characterized by a low level of external threat and a high level of collateral harm; it is located in the lower right quadrant of Figure 1.1. The expansive infrastructure in foreign countries which the U.S. continues to make use of despite the nonexistence of a threat, are often viewed as privileges which accrue to the United States simply as a result of its status. Finally, it is important to point out that the transition from one type of protection regime to another could, at least in theory, move in any direction. In my particular cases, however, I have found that the transformation has occurred in a clockwise direction, based on Figure 1.1.

In sum, I argue that the instability of the U.S. overseas military presence results not only from democratic transitions within the host nation, as others have argued, but from the fact that U.S. protection regimes can vary over time and space. In allied countries, access to facilities depends on the U.S. presence being widely regarded as legitimate, and yet this legitimacy can be called into question for a variety of reasons, including: (1) if the external threat ceases to exist, (2) if the threat against which the U.S. claims to protect was in fact called into being as a result of previous U.S. policies, (3) if the U.S. military itself appears more ominous than the threat against which it claims to protect, or simply because (4) the U.S. military is generally not accountable to those in whose name it is offering protection. The production of violence or concrete harm, although meant to reassure allies and create protection, may not inevitably, but often enough has instead resulted in the production of collateral harm.

Understanding the Consequences and Impact of Anti-Base Movements

The question of why people engage in collective action is perhaps the oldest and most central question in the literature on contentious politics. To answer almost any other question regarding social movements, it is first necessary to understand why they emerge. The first generation of social movement scholars in the 1950s often approached the subject from a Durkheimian perspective in which social cohesion was believed to be a normal state of affairs. Social unrest, therefore, was an exceptional situation that occurred because of the accumulation of grievances which led to a partial breakdown of an otherwise harmonious social life. These types of explanations are often referred to as strain and breakdown theories.[46] By the late 1960s and 1970s, this approach had come under attack by those who argued that grievances are more or less always present, as conflict, rather than cohesion, was intrinsic to society. These scholars, more influenced by Marx than Durkheim, believed that protest emerges when enough resources can be mobilized or sufficient opportunities to protest present themselves. These types of explanations are often referred to under the heading of resource mobilization literature.[47] As scholars turned their attention to studying resources, political opportunity structures, and mobilization strategies, the analysis of grievances fell somewhat by the wayside. Those who returned to the scrutiny of grievances have distinguished between two general types: long-standing grievances, such as poverty or racial or sexual discrimination, and "suddenly imposed grievances," such as oil spills or nuclear accidents.[48]

In the academic divide between strain and breakdown and resource mobilization theorists, I argue for a somewhat modified approach. Grievances regarding the American military presence were neither permanently present

---

[46] Some early strain and breakdown theorists include Park and Burgess 1921, Blumer 1951, Davies 1962, Geschwender 1968, and Gurr 1970.

[47] Scholars who contributed to the critique of strain and breakdown theories and developed the resource mobilization paradigm include McCarthy and Zald 1973, Schwartz 1976, Tilly 1978, and McAdam 1982.

[48] The nuclear accident at Three Mile Island in Pennsylvania is referred to by Walsch (1981) as a "suddenly imposed grievance" which triggered a collective response by the affected communities. Recently, some scholars have modified the original strain and breakdown paradigm; Snow et al. argue that "disruptions of the quotidian" are likely to provoke mobilization. These disruptions, however, usually involve some threat to everyday routines and may therefore also be considered a sudden imposition, Snow et al. 1998.

from the beginning nor did they suddenly emerge and soon disappear. Instead, grievances emerged gradually as the nature of the American military presence changed over time. During the early years of the Cold War, the U.S. military presence in Turkey and Germany provided protection from the Soviet Union. As the threat was real, and the U.S. presence relatively benign, I have termed this period as representing "legitimate protection." In this sense, it was a type of public good, akin to clean air, of which virtually no one disapproved. However, in its endeavor to protect its protégés and threaten its enemies through its ability to inflict concrete harm, the U.S. presence instead caused collateral harm to its allies. Grievances emerged because the U.S. presence was believed to cause harm to those it claimed to protect.

Having schematically outlined why grievances emerge, I briefly address the question of whose grievances are at stake. For analytical purposes, it is useful to distinguish between anti-base and anti-war movements, although at times, of course, the two movements may merge, as in the large wave of protests in Turkey in 2002–2003 against the invasion of Iraq. Anti-war movements are often opposed to a specific military intervention, whereas anti-base movements are opposed to the permanent infrastructure that makes interventions possible. In reality of course, there is some overlap and the lines between the two may become blurred. It is also useful to categorize the goals of anti-base movements as being self-protection, whereas the goals of anti-war movements are often (but not always) the protection of others. These are perhaps the opposite, or at least different, from so-called NIMBY (or "not-in-my-backyard") movements as they are not motivated primarily by self-interest. These differences in the movements are also related to the different types of transnational harm, as discussed previously.

Sources of Vulnerability and Sources of Power
This raises the question of how such movements can be effective when the institution they are protesting is essentially insulated from popular resistance. As outlined earlier, even the most isolated U.S. military bases rely on the civilian infrastructure of host countries to function. This includes roads, harbors, airports, and airspace, but sometimes also private property for maneuver exercises or local businesses for the provision of goods and services. In some countries, U.S. officials may rely on the local police to protect installations or guard checkpoints at the entrances and exits. The use of military facilities may at times be subject to parliamentary scrutiny. Finally, because not all tasks can be performed by U.S. soldiers or sailors, any number of local civilians may be hired to work on the bases. It is this dependence on the civilian infrastructure, parliamentary approval, and labor power that makes the U.S. presence potentially vulnerable to social unrest. It is precisely these sources of vulnerability that are potential sources of power for the otherwise disenfranchised protectariat. However, they are of course only *potential* sources of power: to translate into actual sources of power, they must be acted on, which brings us to the question of tactics.

Within the theoretically unlimited repertoire of protest forms, scholars have identified a number of tactics utilized by social movements; four of them are briefly discussed. In 1969, Ted Gurr maintained that violence is the basic resource available to social movements.[49] Looking back at the unrest of the late 1960s and early 1970s in many American cities, Frances Piven and Richard Cloward contended that it was not necessarily violence, but rather disruption that is often the most effective tactic for those who are excluded from institutions. With few other forms of leverage, the marginalized can disrupt the rhythms of everyday life through spontaneous acts of disobedience.[50] James De Nardo, on the contrary, believes that mobilization of large numbers of people is the most effective tool available to the masses.[51] Finally, Sidney Tarrow argues that novelty is important because when new or unusual protest forms are implemented, the authorities are often caught off-guard and cannot respond either immediately or effectively.[52]

More recently, research has shifted away from the study of tactics to the larger environment in which unrest takes place. This research has taken on two trajectories: those who study the role of public opinion and those who study the political context more generally. For the latter trajectory, the concept of "political opportunity structure" has become influential in analyzing two factors: the structure of the state and the wider political landscape including alliance networks and oppositional groups. My own critique of the literature is that, although the study of external variables clearly enriches the previous literature that was more focused on the internal variables of social movements, both tend to place social unrest within the context of nation-states. Although many movement claims are directed to the state, even a general definition of social movements allows that this may not always be the case. For Tarrow, social movements are "sustained challenges to powerholders in the name of a disadvantaged population living under the jurisdiction or influence of those powerholders"[53] Clearly, not all holders of power are democratically elected to represent the constituents of a nation-state. The definition of contentious politics as involving "interactions in which actors make claims bearing on someone else's interests, leading to coordinated efforts on behalf of shared interests or programs, in which governments are involved as targets, initiators of claims, or third parties" also allows for the possibility that the government which is targeted may not have jurisdiction over those making claims.[54]

This raises the question of how to determine outcomes of social movements when those movements are not directed against the nation-state, but a foreign

---

[49] Gurr 1970.
[50] Piven and Cloward 1977.
[51] De Nardo 1985.
[52] Tarrow 1989.
[53] Tarrow 1996: 874.
[54] Tilly and Tarrow 2007: 4.

military presence. Here it becomes necessary to define what would be considered a successful outcome of an anti-base movement. For those movements
advocating withdrawal, the decision to leave is clearly a "success" from the
point of view of anti-base activists. However, the decision to stay in the face of
opposition does not necessarily represent a complete failure for the movement.
Think tank reports can provide insights in this regard.

Short of evicting a military base, various forms of access restrictions can
also pose problems for commanders in the field. These two types of problems are defined in a report entitled "Meeting the Anti-Access and Area-Denial
Challenge." "If anti-access (A2) strategies aim to prevent US forces entry into
a theater of operations, then area-denial (AD) operations aim to prevent their
freedom of action in the more narrow confines of the area under an enemy's
direct control."[55] If so-called A2 or AD operations are carried out by enemy
forces, it is possible to deflect this challenge through the use of superior technology. For example, the report goes on to describe how the Air Force's Global
Strike Task Force (GSTF) can "kick in the door to denied airspace by taking out
advanced surface-to-air missiles (SAMs) as well as critical mobile targets such
as enemy mobile-missile launchers."[56] What is to be done, however, if airspace
is denied by an allied host country? In an article by Lieutenant Sean Atkins
entitled "Unwanted Allies: What Influences Negative Domestic Reactions to
Deploying Forces into Allied States?" published in *Air Power History*, the
author discusses protests in Britain against the base at Greenham Common,
which continued for nineteen years, from 1981 until 2000, and were organized
by British women. Atkins describes how the U.S. Air Force had no protocol
for how to deal with the protests: "Though the US Air Force was prepared for
nuclear strikes, it had not planned on attacks based in the field next door."[57] He
goes on to describe the impact of the resistance by local citizens:

Instead of sneaking onto an armed military installation, protestors could simply sit
down at the gates, effectively blocking the cruise convoys from exiting and preventing
their dispersal.... They would offer passive resistance for as long as possible, thus grind-
ing most base activity to a halt until they were eventually removed.[58]

According to Major General Fox Conner, who was perhaps best known
for his mentorship of Dwight Eisenhower when he was still a young army
officer: "Dealing with the enemy is a simple and straightforward matter when
contrasted with securing close cooperation with an ally."[59] The example of
Greenham Common would seem to illustrate this seemingly paradoxical situation, as most Cold War military doctrines were based on a binary logic of
friends and enemies in which it was assumed that one's allies will necessarily

[55] Krepinevich 2003.
[56] Ibid: iii.
[57] Atkins 2004: 40–51.
[58] Ibid: 49.
[59] Cited in Huston 1988: 128.

have the same interests as one's own. The emergence of anti-base opposition in allied countries gave the lie to these assumptions and underlined the tripartite nature of the Cold War conflict.

Within the context of this project, each of the groups I study had different goals and employed different tactics. Of the four basic tactics discussed earlier, all of them were implemented, albeit with varying degrees of effectiveness. Short of ousting the U.S. military, it is possible to achieve a number of other things including, but not limited to: creating access restrictions or temporary access denials to either territory, airspace, or certain facilities; disrupting access to infrastructure; implementing or enforcing environmental standards that make certain military activities too costly or difficult to conduct; preventing military expansion or construction projects; shutting down base operations through strikes of civilian base employees; creating shortages by refusing to supply the base with goods and services; and making the U.S. military accountable for its actions. The common denominator of these potential goals is that they limit the autonomy of the military. Based on these considerations, I argue that the protectariat possesses several sources of social power:

- Disruptive Power: This includes the ability to disrupt military activities, for example, by blocking access to various parts of the off-base infrastructure on which the U.S. military relies, such as roads, airports, and harbors.[60]
- Power of Access Denial: This includes the ability to deny or restrict access to territory, airspace, or facilities.
- Power of Passive Resistance: This includes the ability to engage in various types of noncooperation, such as refusing to supply the base with goods and services.
- Power of Armed Resistance: This refers to the ability to create a hostile or insecure environment through violent attacks or kidnappings of military personnel.
- Structural Power: This includes the ability to strike or engage in other forms of labor unrest. Whereas the first four types of social power can potentially be utilized by anyone belonging to the protectariat, I argue that structural power accrues only to those who are hired to work on the bases, as the protection regime relies on their labor power in order to function.[61]

As shown in the following chapters, these different types of social power may be exercised in different ways and do not always correlate with specific tactics.[62] If these various types of unrest do not succeed in creating access restrictions or

---

[60] As discussed previously, Piven and Cloward are credited with having introduced the idea of disruptive power. See Piven and Cloward 1977.

[61] Erik Olin Wright distinguishes between associational power, which workers possess when they form associations such as trade unions or political parties, and structural power, which workers possess by virtue of their location within the economic system. Wright 2000: 962.

[62] A table summarizing the types of social power and types of protest is in Chapter 5.

otherwise disrupting military activities, I define this type of protection regime as a *functional protection regime*. The question of functionality does not refer to the raison d'être of the U.S. presence (i.e., whether it serves the purpose of protecting its protégés or simply serves U.S. interests as a platform for power projection), but instead refers simply to its ability to operate. If base operations are significantly affected by social unrest or shut down altogether, I argue that the protection regime has become a *dysfunctional protection regime*. To clarify, it may be possible for hundreds of thousands of people to take to the streets in protest against the U.S. military presence, but for base operations to remain functional. On the other hand, it may be that the local population largely welcomes the U.S. presence, but for a small protest event to significantly disrupt base operations. What is decisive for this study is not the level of popularity of the U.S. presence, but its functionality – not the size of the protest, but its ability to render the U.S. presence dysfunctional.

Having defined what a successful outcome of anti-base unrest might be, the question of establishing causality remains to be solved. Because I am interested in how the U.S. military responds to protest events, using government documents allows me to establish with a fairly high degree of certainty the outcome of the unrest, in terms of its impact on the U.S. presence. Relying on archival documents is both easier and more difficult than other methods of tracing movement impact. The trials and tribulations of making one's way within the labyrinth of state archives, in which each branch of the government has its own archives and its own rules for access and declassification and categorization of documents, could constitute a separate chapter. Furthermore, beginning in 2003, the Bush Administration proceeded to slash the funding for state archives while at the same time ordering the reclassification of previously declassified documents, making the already glacial process of obtaining documents through the Freedom of Information Act even slower.[63] Once these previously classified documents have been released to the public (if often in redacted form), however, one can be fairly sure that the information made available is more reliable than journalistic accounts, interviews, memoirs, or other material. Because the documents were meant for internal use, as a way of communicating between various branches of the government and between those located in Washington, DC and those "in the field", the reports are often candid and unflinching. It is not uncommon to

[63] On March 25, 2003, President Bush signed Executive Order 13292 which replaced the Clinton-era Executive Order that required automatic declassification of federal records after twenty-five years, with the exception of certain categories of documents. The government is now able to keep certain records classified indefinitely, and has vested the vice president with the power of classifying documents. It has also expanded the category of documents that are exempt from automatic declassification to include information that could "impair relations between the United States and a foreign government." See the Society of American Archivists' Web site, "Bush Issues New Secrecy Executive Order": http://www.archivists.org/news/secrecyorder.asp.

find causal statements linking descriptions of protest events and suggestions or orders on how to respond.

In sum, within the context of this study, I argue that the emergence of widespread social unrest in allied host countries will be an indicator that the protection regime the U.S. military presence embodies has shifted from one of legitimate protection to one that can be characterized as pernicious protection. If U.S. military activities are not negatively affected by the protest, I argue that the U.S. presence is a functioning protection regime. If, however, the social unrest succeeds in implementing anti-access or area-denial strategies or more generally limiting the autonomy of the U.S. military, I argue that the U.S. presence may be approaching a dysfunctional protection regime.

CASE SELECTION

As mentioned, fourteen European countries hosted U.S. troops and bases during the Cold War. Faced with such a large array of potential research sites, it was necessary to make some choices. Preliminary fieldwork suggested that Germany and Turkey were ideal countries in which to conduct in-depth case studies into the evolution of opposition to the U.S. presence because both countries have hosted U.S. troops and bases for more than half a century and both continue to be pivotal within NATO. Both countries received military aid via the Marshall Plan and were characterized as "security dependent" during the Cold War as both relied on U.S. protection. Furthermore, the political landscapes of both countries feature multiparty political systems with strong Centre-Right (CDU in Germany, and DP [later AP] in Turkey) and Centre-Left parties (SPD in Germany, and CHP in Turkey)[64] as well as small parties on the left that were involved in campaigns against NATO and the U.S. presence (the Greens in Germany in the 1980s, and the Labor Party in Turkey in the 1960s).[65] At different times and for different reasons, these small parties joined forces with other groups in society to oppose either certain aspects of the U.S. military presence or question its very legitimacy, leading to a changed consensus on the U.S.-sponsored security umbrella. These cycles of contentious politics led to or were accompanied by crises in the bilateral relationships. For the purpose of this study, a crisis will be defined as an exceptional situation in which the political order is called into question. A true crisis is more than a momentary clash of personalities among diplomats or heads of state; it involves the shifting of allegiances among larger societal forces that are not controlled by a single person or office. In what follows, I give a brief overview of the movements that emerged in opposition to the U.S. military presence.

[64] Please refer to the List of Acronyms.
[65] CDU: Christian Democrat Union; DP: Democratic Party; AP: Justice Party; SPD: Social Democratic Party; CHP: Republican People's Party.

The Turkish Case

In many ways, Turkey represented an ideal country to host U.S. forces. As the successor state to the Ottoman Empire, the Turkish Republic initially welcomed the creation of an American presence. Having never been colonized, it lacked a deeply rooted tradition of antipathy toward the West, but, on the contrary, had begun a program of economic and political Westernization since the founding of the Republic under Atatürk. Furthermore, the U.S. presence in Turkey was established in reaction to Soviet demands for joint control of the straits, hence the United States was initially viewed as a benevolent hegemon offering legitimate protection against an outside aggressor. The early phase of the U.S. presence in Turkey therefore belongs in the upper left quadrant in Figure 1.2. Under these circumstances, the U.S. presence expanded to include 30,000 troops at its height in the mid-1960s. This was comparable to the size of the U.S. presence in the United Kingdom or South Korea, although obviously much smaller than the U.S. presence in West Germany.

However, a deterioration of bilateral relations began to set in around 1964, initiating the most severe crisis of U.S.-Turkish relations during the Cold War. Until now, the literature has explained the growing tensions as being largely caused by the Cyprus issue. Many scholars of U.S.-Turkish relations have rehearsed the story of how bilateral relations began to deteriorate when President Johnson warned Prime Minister Inönü not to use U.S.-supplied weapons to intervene in Cyprus in 1964.[66] In other words, the crisis has been analyzed as a conflict of national interest between Washington and Ankara regarding a third entity. However, I demonstrate that the unrest in Turkey was triggered by an accumulation of tensions relating to the U.S. presence in Turkey, including the Cuban missile crisis,[67] the U-2 spy plane incident, and other reconnaissance flights that had taken off from bases in Turkey without Ankara's approval; the use of Istanbul, Izmir, and other port cities for rest and relaxation during visits of the Sixth Fleet; the perceived disrespect toward Turkish citizens, in particular women; and labor practices involving Turkish employees at the joint defense facilities.

Because of these events, many Turkish citizens began to think that the U.S. presence was doing more harm than good; here the protection regime shifts to what I have termed pernicious protection, located in the upper right quadrant of Figure 1.2. Beginning in the mid-1960s, the Türkiye İşçi Partisi (TIP) under the leadership of Mehmet Ali Aybar began to mobilize support for their campaign aimed to expel the U.S. bases, which they saw as an infringement of Turkish sovereignty. Aybar and other party members described their

[66] Harris 1972; Kuniholm 1985; Candar 2000; Gueney 2005.
[67] As I discuss in more Chapter 2, the Soviet Union had deployed missiles in Cuba in response to the Jupiter missiles that the United States had previously stationed in Turkey.

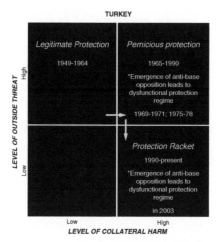

FIGURE I.2. Turkey protection graph.

struggle as a second national liberation movement, after the 1919–1923 War of Independence. In addition to opposition within the parliament, protest against the U.S. presence took other forms as well, including strikes of Turkish workers at American military facilities in Ankara and Incirlik. Base employees were organized in the Harb-İş union and were highly politicized, as they scheduled their strikes to coincide with NATO maneuvers so as to maximize their disruptive potential. The protests that received the most media attention were certainly those against visits of the Sixth Fleet to Istanbul and other port cities, whereas the kidnappings of allied servicemen by the Turkish People's Liberation Army (THKP-O) were certainly the most violent. Finally, there were various outbreaks of unorganized violence and rioting, such as the burning of Ambassador Komer's car during his visit to the Middle East Technical University in 1969. Another spontaneous eruption happened in November 1966 when a group of American G.I.s were accused of harassing Turkish women; almost immediately a mob gathered and rioting continued for three days, involving more than 2,000 people.

Were these events triggered by the dispute over Cyprus, as much of the literature suggests? I argue that these contentious politics had more to do with an accumulation of tensions relating to the U.S. presence *in* Turkey, which many were beginning to perceive as increasingly pernicious in nature.

As I argue in more detail in Chapter 2, the transformation from legitimate protection to pernicious protection and the ensuing challenges to the American military presence began in Turkey before the Vietnam War had become a quagmire and before the emergence of an international anti-war movement. In Germany, as explained later, the transformation from legitimate protection to pernicious protection occurred after the end of the Vietnam Conflict. In other words, in neither case did the Vietnam War trigger anti-base sentiment. It would appear that whether a host nation accepts or contests the U.S. military presence depends more on the legitimacy of the U.S. presence in the local context than within the global landscape.

The German Case

Most of the literature on the U.S. presence in Germany focuses on the early period, from the mid-1940s to the mid-1950s.[68] However, the U.S. presence was not a static entity, but underwent dramatic changes since the original occupation period. From May 8, 1945 until May 5, 1955, the American military presence in Germany took the form of a *Besatzungsmacht*, or occupying power, along with the French and British forces in the West and the Soviet army in the East. Following the conceptual framework discussed earlier, the United States and its Allies had defeated the Nazi regime and taken over the administration of Germany. During most of this period, the protection regime was characterized by a low level of outside threat and a relatively high level of legitimacy of the U.S. presence, as there was little organized resistance after the Wehrmacht was defeated; in Figure 1.3 at the end of the introduction, this period of the U.S. presence is therefore in the lower left quadrant. Beginning in the late 1940s, however, the anti-fascist coalition between the USSR and the three Western Allies was breaking down and the Cold War was beginning to emerge. By the mid-1950s, the U.S. military was referred to as a *Schutzmacht*, or protective power, by the German people. The U.S. presence had been transformed from an occupation to legitimate protection, the Germans had been given a new and upstanding identity based on anti-communism, and the Soviet Union had become the common enemy. This period is therefore located in the upper left quadrant of Figure 1.1 (legitimate protection), as it is characterized by a high level of external threat and a high level of U.S. credibility as an effective protection regime.

Another dramatic transformation of the American protection regime occurred during the last decade of the Cold War. Several decades of nearly seamless cooperation between the two allies led not only theorists of democratization but also basing strategists to consider German-U.S. cooperation as a major success story of the postwar period. This was all the more remarkable because West Germany hosted the largest U.S. presence of any country in the world: on average, 250,000 troops were stationed there throughout the Cold War, with about 250,000 dependents in addition to the soldiers. However, the NATO double-track decision in 1979 to station intermediate-range nuclear missiles in Germany and four other NATO allies led to a serious deterioration of bilateral relations. This decision is generally considered as having triggered

[68] Studies that concentrate on the early years of the American presence include: Glaser 1946; Davidson 1959; Gimbel 1968; Hillel 1981; Höhn 2002; and Goedde 2003.

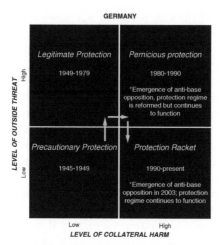

FIGURE 1.3. Germany protection graph.
Source: http://www.heritage.org/Research/NationalSecurity/cda04–11.cfm

the largest social movement in postwar German history, but also led to the fall from power of the Social Democratic Party (SPD) in 1982 and helped to catapult the Green Party from local politics into the Bundestag. The impact of this upheaval on domestic German politics has been widely documented: security policies ceased to be the preserve of the elite but became part of the wider public discourse, and the consensus on nuclear issues changed forever.[69] However, the problems that arose during the last decade of the Cold War had major ramifications not just for domestic German politics, but also for the continuing U.S. military presence there.

The first half of the 1980s was defined by the widespread anti-nuclear protests.[70] These movements spread to a much larger portion of the German public than the youth radicalism of the 1960s, which was generally limited to the student-age population in major urban centers. The anti-nuclear and environmental movements of the 1980s branched out into both urban and rural areas, including much of the middle class, and eventually found support from major church organizations, professional associations, and trade unions. In addition to opposition to the missile deployment, the critique of the U.S. presence in Germany included demands by citizens' initiatives (*Bürgerinitiativen*) to reduce the level of noise pollution caused by low-flying military jets, to decreasing the damage caused by maneuvers such as the REFORGER exercises, to demands that true peace could only be attained when all foreign troops withdraw and Germany regains its full sovereignty. In other words, the peace

[69] Cooper 1996; Breyman 2001; Rohrschneider 1993.
[70] Spoo 1989; Angerer 1990; Achilles 1987; Lafontaine 1983, Mechtersheimer 1984.

movement's critique of the U.S. presence was more far-reaching than is generally acknowledged in the English-language literature and represented a major
shift in public perceptions of the U.S. presence in Germany.

In addition to the anti-nuclear movement and the Green Party, the U.S. military presence faced two other problems: labor disputes on the bases (the U.S.
military was the eighth largest employer in West Germany during the Cold
War) and violent attacks by terrorist groups. Between the early 1970s and the
mid-1980s, the Rote Armee Fraktion (RAF) and Revolutionäre Zellen (RZ)
carried out dozens of attacks against U.S. military facilities and personnel.

During this period, many people began to call into question whether the
American military presence still served the purpose of protecting Germany,
as many local citizens felt that they were in fact being taken hostage by the
Americans as they were stationing their most deadly first-strike weapons in
densely populated areas of Western Germany. If the Soviets tried to attack
the nuclear weapons depots, a substantial part of the civilian population of
Germany would be endangered, which would have necessitated retaliation,
turning Europe into a nuclear battlefield. Because Reagan talked about the possibility of limited nuclear war in Europe, many Germans began to feel that the
U.S.-controlled nuclear weapons – and especially the administration's apparent
willingness to fight a nuclear war – represented a greater threat than the Soviet
Union and began to use words such as *Erpresser* (or "racketeer") instead of
Schutzmacht to refer to the U.S. presence. I argue that, by the early 1980s,
the U.S. presence in Germany was widely perceived as a pernicious protection
regime, characterized by a continuing external Soviet threat but a low level of
U.S. legitimacy.

Comparing Cases

In some ways these two countries represent two broader types of anti-base
movements, which until now have been studied separately, with Germany representing the anti-nuclear peace movements of Northern Europe and Turkey
representing the independence or more avowedly anti-imperialist movements
of Southern Europe. Whereas in Germany opposition movements stemmed
from a growing environmental and anti-nuclear awareness that was rooted
in the pacification of Germany after World War II, the U.S. presence in Turkey
was seen as preventing the full exercise of Turkish sovereignty, including its
ability to intervene in Cyprus. It is important, however, not to over-emphasize
cultural and socioeconomic differences because, from the perspective of the
Pentagon, the two situations were indeed similar because they both represented
increasing political and economic costs as well as an accumulation of access
restrictions. This meant that the force structure in both countries was becoming more expensive and yet less effective.

Although low-level unrest occurred during the entire Cold War period
in both West Germany and Turkey, I focus on the high point of activity in
each country as a coherent protest cycle. This occurred from 1964 to 1975 in

Turkey, with the cycle climaxing between 1968 and 1971. The protest cycle in Germany spans the period between 1979 and 1987, with the high point of activity between 1981 and 1984. Despite cultural, socioeconomic, and religious differences, which may explain why the local population challenged different aspects of the U.S. presence, opposition to the U.S. presence in West Germany and Turkey displayed similar patterns. Each country developed: (1) parliamentary protest with a small political party spearheading a campaign against the U.S. presence and/or NATO (the Greens in Germany and the Labor Party in Turkey), (2) non-violent civil disobedience in the streets (blockades of missile deployments in Germany and disruptions of Sixth Fleet visits in Turkey), (3) labor unrest involving German or Turkish civilian employees on American bases, and (4) violent attacks by terrorist groups (the RAF and RZ in Germany and the THKP-C and THKO in Turkey). Taken together, these various types of unrest form an accumulation of contentious politics rather than a single coherent social movement.[71] Whereas social scientists may be tempted to analyze these social movement actors separately, from the perspective of U.S. diplomats and military commanders, the various forms of contention were categorized as "problems with the host country" or "host nation activities." At numerous times, even the U.S. ambassador to Germany or Turkey, members of the U.S. Congress, and high ranking administration officials such as Henry Kissinger became involved in negotiating resolutions to the problems, further evidence that maintaining a functioning military presence was a high priority. How this happened is discussed in detail in the chapters that follow.

For various reasons, sociological research on social movements in these two countries has served to obscure rather than elucidate these similar patterns of protest. For this reason, further research along these lines is necessary. Obviously, the deterioration of bilateral relations in both countries was not merely a clash of elite interests, but was a crisis "from below," or a societal upheaval that revealed profound differences in how Americans and their allies perceived the security landscape. Although the impact of this upheaval on domestic German and Turkish politics has been documented (albeit with varying degrees of thoroughness), the ramifications for the U.S. military presence have been neglected by social scientists, despite the fact that many of the contentious issues that arose during the Cold War still persist today. Finally, despite the similar trajectories of Turkish and German anti-base protest, they did not result in similar impacts on the U.S. presence in the respective country. In fact, the U.S. military responded quite differently to the protest in each country.

In Germany, the United States was obliged to make some concessions to movement demands by adhering to increased environmental standards, ending the practice of flying military aircraft at altitudes as low as seventy-five meters and reducing the property damage caused by NATO maneuvers. However,

---

[71] Tilly and Tarrow: 2006.

TABLE 1.1. *Types of Anti-Base Unrest in Turkey and Germany*

	Protest in the Parliament	Nonviolent Civil Disobedience	Violent Attacks	Labor Unrest on Base
Turkey	Turkish Labor Party (TIP)	Sixth Fleet protests	THKP-C THKO	Strikes scheduled during NATO maneuvers
Germany	Green Party	Blockades at missile depots Maneuver disruptions	RAF RZ	Symbolic work stoppages

despite the fact that anti-nuclear demonstrations in 1981 and 1983 drew the largest crowds in postwar German history with more than a million participants, the United States was not deterred in its decision to deploy the Pershing II and Cruise missiles. After deployment, despite daily blockades at missile depots such as Mutlangen and Neu-Ulm that continued for almost four years and that sometimes drew prominent citizens such as Heinrich Böll, Erhard Eppler, and Günter Grass, the U.S. Army was not deterred in its decision to continue to rotate the missiles through these small villages on a regular basis.

In Turkey, the response was quite different. After several protests against visits of the Sixth Fleet, which attracted at most 30,000 people and were nowhere near the size of the mass rallies in Germany and which did not attract prominent members of the Turkish intelligentsia, the navy decided in 1969 to stop making port calls to Turkish cities. Furthermore, the United States withdrew more than 20,000 troops (75 percent of the total) and forty of the fifty-four bilateral agreements concerning the U.S. presence had been modified to meet Ankara's demands. Curiously, neither the literature on U.S.-Turkish relations nor the social movements literature has provided any explanation for why the United States suddenly withdrew the majority of its troops from Turkey (Tables 1.1 and 1.2). What can explain the different U.S. response? What were the larger implications of anti-base unrest in both countries?

Over the course of the protest cycles in Germany and Turkey that I study, all four types of protest tactics were utilized and are discussed at greater length in the following chapters. A more difficult question to answer is the issue of how the U.S. authorities perceived and responded to these various protest events. As discussed under "Data Sources" at the end of this chapter, I have reviewed thousands of pages of recently declassified archival documents in order to answer this question. Based on the archival data, it is clear that not all forms of protest were equally effective. Although the mass character of the German peace movement was certainly significant in terms of its repercussions for the German political scene, it did not appear to increase its impact on the U.S. presence, nor did the novelty of having prominent citizens partake in protest events. The violent attacks of terrorist groups in both countries also did not have a

TABLE 1.2. *Consequences of Anti-Base Unrest during the Cold War*

	Event most often cited as triggering protest cycle	Factors in the transformation from legitimate to pernicious protection (Collateral harm)	Consequence of unrest – Domestic response	Consequence of unrest – U.S. response
Turkey	Johnson letter in 1964	Use of Istanbul and Izmir for R&R Unfair labor practices on base U-2 spy plane incident Infringements of Turkish sovereignty	PM Ecevit (CHP) closes U.S. bases 1975–1978	The United States withdraws 75% of its personnel; navy stops making port visits after 1969; modifications in a bilateral treaty
Germany	NATO double track decision in 1979	Property damage during maneuvers Low-flying jets Missile deployment Creation of new bases in protected habitats	SPD falls from power; joins anti-nuclear cause Green party elected to the Bundestag	The United States obeys environmental standards and reduces maneuver damage The United States eventually removes missiles after the INF Treaty in 1987

lasting impact on the U.S. military because these groups could be easily criminalized and marginalized. Although perhaps counter-intuitive, violent attacks against U.S. facilities or personnel were in some ways the easiest to respond to because they legitimized an increase in police repression and merely required some "technical" adjustment to an already existing strategy. By contrast, those protest events which were apparently the most difficult to counter were those which were focused on disrupting the normal functioning of the military presence through non-violent means. Although disruptive tactics were utilized in both cases (blockades of missile rotations in Germany and disturbances of Sixth Fleet visits in Turkey) the impact of labor unrest on the military bases themselves was perhaps the most disruptive – and therefore the most disturbing – of all. In Germany, the trade unions participated in symbolic acts such as five-minute work stoppages that were organized by the peace movement.

However, during the entire Cold War period and even during the height of
peace movement activity in Germany, the strikes that occurred never seriously
threatened the functioning of any of the U.S. bases. In Turkey, on the contrary,
base workers not only went on strike for weeks at a time, but even scheduled
their strikes to take place during NATO maneuvers. As "critical cogs in the war
machine" base workers were able to shut down base operations without being
criminalized or marginalized.

ANTI-BASE UNREST IN THE POST–COLD WAR ERA

At a more general level, anti-base unrest in numerous regions of the world,
including an increase of access denials and access restrictions, led the Pentagon
to reconsider its global force posture at the end of the Cold War. Although the
Revolution in Military Affairs has been discussed within the context of new
weapons systems, changing national interests, and the unipolar world order,
it is also a response to the social problems that emerged during the Cold War.
In 1997, Secretary of Defense Cohen appointed the National Defense Panel
"to provide an assessment of alternative force structures for the U.S. military
through the year 2010."[72] One outcome of this decision was the Global Defense
Posture Review (GDPR). The GDPR represents the most fundamental transformation of the U.S. basing posture since World War II. The GDPR planned
to reduce U.S. forces in several major base hosts and establish a network of
smaller, more flexible facilities deep inside the Eurasian landmass, and in Africa
and Latin America. One explicitly political goal of the GDPR is to reduce the
footprint and local friction caused by the U.S. military presence by establishing
smaller facilities of a less permanent nature that will be less politically controversial and socially intrusive within host countries.[73] Ironically, it would seem
that the basing literature had in the end agreed with the more critical anthropological literature that fewer bases would mean fewer problems. The recent
resurgence of anti-base unrest in Turkey is an indication, however, that these
assumptions are mistaken.

In 2003, the United States had a skeleton crew of some 2,000 permanent
troops in Turkey, the smallest presence since the late 1940s. When the Bush
Administration began drawing up plans for the invasion of Iraq, they looked
to Ankara. To open up a northern front, the use of Incirlik Air Base as well
as other facilities in Turkey was considered crucial. From an economic standpoint, Turkey was still recovering from the financial crisis in February 2002,
one of the worst in its history. To ensure Turkish cooperation, Washington

[72] http://www.dtic.mil/ndp/.
[73] As David Isenberg points out, only some of the new facilities will be smaller, those which were
built in Iraq were intended to be just as large as those in Germany. Isenberg: "The US Global
Posture Review: Reshaping America's Global Military Footprint," 2004: http://www.basicint.
org/pubs/Notes/BN041119.htm.

offered Ankara an aid and credit package amounting to $26 billion and even agreed to have its soldiers tried in Turkish courts if they violated Turkish law outside the bases.[74] That the Pentagon was willing to make such a concession underlines Turkey's importance for the war planners.[75] On March 1, 2003, however, the Turkish parliament voted "no" to the deployment of U.S. forces in Turkey and the use of Turkish territory for the invasion of Iraq, succumbing to the anti-war sentiment in the population.[76] Although Berlin allowed the United States to use bases in Germany for the war, the German financial contribution to the Iraq War effort was minimal.[77] The ruling on June 21, 2005 by the German Federal Administrative Court (*Bundesverwaltungsgericht*) that the Iraq War and Germany's support for it was cause for "grave concerns in terms of international law" was in several ways unprecedented.[78] Secondly, the attempt to prosecute former Defense Minister Rumsfeld by German attorney Wolfgang Kaleck received much attention and generated ill will, including suggestions by various observers that Washington should threaten to close down the remaining facilities in Germany.[79]

As the German and Turkish case studies illustrate, even NATO allies cannot guarantee "assured access." Whether non-NATO allies in Eastern Europe, Central Asia, Africa, or the Middle East will offer a more supportive environment for U.S. operations is an open question. Of course, imperial structures are complex phenomena subject to both internal and external forces. Michael Doyle has argued that the dynamics of the international system can explain when or why imperial expansion has accelerated, whereas the needs and interests of the metropole largely determined where in the periphery this expansion took place. Finally, whether direct or indirect rule evolved depended most on the attributes of the peripheral society.[80] I argue that, although the process of losing overseas territories is a complicated and multifaceted process, the principal source of pressure to change the status quo arises in the nations that host U.S. bases.

I argue that anti-base movements are best understood as analogous to anti-colonial movements. Whereas anti-colonial movements demanded decolonization, anti-base movements advocated at a basic level for some form of demilitarization. Whereas anti-colonial movements demanded the democratization of state-making, anti-base movements demanded the democratization of war-making.

[74] Yetkin 2004.
[75] "Holding Its Ground," *Newsweek*, January 20, 2003.
[76] Altınay and Holmes 2009.
[77] CRS Report, "Post-War Iraq: A Table and Chronology of Contributions," March 18, 2005.
[78] Schultz, Nikolaus, "Was the war on Iraq illegal? The German Federal Administrative Court's Ruling of 21st June 2005" in *German Law Journal*, Vol. 7, No. 1, 2006.
[79] Rabkin, Jeremy, "Rumsfeld Accused; anyone home in the State Department?" *Weekly Standard*, December 4, 2006.
[80] Doyle 1986.

In the following chapters, I illustrate how this process played out differently in Germany and Turkey. However, the inherent paradox in which the United States finds itself in Germany and Turkey is common to every potential host country. At the same time as the United States claims to favor the spreading of democracy, it has established and continues to rely on a global network of military bases which largely operate outside the realm of the democratic polity of the countries in which they are located. Because U.S. military officials are not directly accountable to the non-U.S. citizens in these countries, we can conceive of them as a "protectariat" who are at the same time citizens of their nation-states as they are subjects of the American empire. And yet, through their mobilization in anti-base movements, they may succeed in disrupting or delaying U.S. military operations and hence the baseworld itself may be made vulnerable to the will of the people in those countries which host U.S. bases. This could entail, in the end, a democratization of the American empire of bases.

DATA SOURCES

Neither broad generalizations based on anecdotal observations nor surveys of changes in public opinion can provide a true understanding of the causes and consequences of anti-base unrest. Now that the Cold War has ended and many of the relevant archives have been opened, it should be possible to illuminate the contention within the NATO alliance and understand the various forms of opposition to the U.S. military presence within European host nations. Because of the dearth of scholarship on anti-base movements in Turkey and Germany, it has been necessary to gather my own data. I have utilized five types of data: (1) recently declassified documents from archives in the United States, including the National Archives and Records Administration (NARA), the Air Force Archives, the Naval Historical Research Agency, the U.S. Army Center for Military History, CIA records, and the National Security Archives (these documents include internal State Department and CIA memos as well as reports by each branch of the armed services that was stationed in Turkey and Germany),[81] (2) programmatic documents from political parties, social movements, and groups engaged in armed struggle in Germany and Turkey,[82] (3) American newspaper archives, (4) German and Turkish newspaper archives,[83] and (5) semi-structured interviews with more than 100 people in the United States,

---

[81] In April 2006, I applied for a Freedom of Information Act (FOIA) for numerous boxes of documents to be declassified. In addition, I have reviewed previously declassified documents that allow me to understand how the U.S. authorities perceived the protests.

[82] I have collected these documents that include brochures, flyers, and pamphlets from archives in Turkey, Germany, and the Netherlands. These documents provide necessary information about the strategies, tactics, and goals of the activists.

[83] Newspaper articles from local papers have allowed me to gain in-depth information about specific protest events.

Germany, and Turkey over a period of several years. I have been committed to interviewing a variety of people so as to understand different perspectives on this topic. I have interviewed military officers, journalists, academic experts, activists, and high-ranking officials in Ankara, Berlin, and Washington.

To explain how the United States perceived and responded to the social unrest directed against the U.S. military presence in each country, it is necessary to rely more on archival material from American than European sources. Some criticisms have been raised about the use of archival material – in particular official documents from state archives – for the research and writing of social history. Official documents may only register issues that were of concern to the state such as diplomatic ties or elite interests, while ignoring the lifeworld (*Lebenswelt*) of the subordinate classes. Within the context of Turkish historiography, Halil Berktay and Şerif Mardin have raised this criticism.[84] Among German historians, Georg G. Iggers has made a similar critique.[85] Within the context of my own research, I am certain that many of the various groups I discuss neither thought of themselves collectively nor acted as a coherent whole, but were perhaps even competing within the social movement sector. And yet, collectively, they posed a challenge to the U.S. commanders in Germany and Turkey, which made their actions of paramount interest to the state. Not only were their voices heard and their actions recorded, but one may be surprised at the meticulous manner in which the history they helped create was recorded, albeit by state officials. Nevertheless, I do not necessarily assume that these documents accurately reflect the true consciousness of those engaged in opposition to the U.S. presence. Instead, I assume that the government documents accurately reflect the process of decision-making regarding the U.S. response to the social unrest and how it affected U.S. basing strategies, which is my primary concern. These official documents have of course been supplemented with materials produced by the social movement activists themselves. And I have benefited enormously from former activists who were invariably the most willing to spend hours of their time retelling the history of the struggles they were engaged in. State officials or military commanders tended to remember less, and were less talkative when they did. The official documents were therefore all the more valuable. Whereas the movement documents were written often with the intention of being as widely disseminated as possible so as to call attention to their cause, state documents were almost always classified and were certainly not produced for public distribution. Therefore, bringing to light materials that have been buried for decades in archives scattered across the United States and Europe may in some sense privilege the state, but the state has two faces, one looking out and one looking in.

---

[84] Mardin 1997; Berktay 1992.
[85] Iggers 2005.

2

Social Unrest and the American Military
Presence in Turkey during the Cold War

INTRODUCTION

In this chapter, I trace the history of the American military presence in Turkey from its rather inconspicuous beginnings in the 1940s until the end of the Cold War. After discussing the expansion of the U.S. presence during the 1950s, I focus on the second half of the 1960s as the high point of social unrest against the U.S. presence during the Cold War. This period witnessed a major shift in public perceptions of the United States and the emergence of social movements that eventually led to major changes in the bilateral relationship between the United States and Turkey, including a 75 percent reduction in the number of U.S. troops stationed in Turkey. Only after the military coup in 1980 and the signing of a new bilateral agreement was it possible to reestablish the U.S. "foothold" in Anatolia. Finally, I show how Turkey's strategic importance for the United States increased with the end of the Cold War and renewed engagement in the Middle East, beginning with the Gulf War in 1991. In terms of my conceptual framework, I show how the legitimate protection regime during the 1950s evolved into a pernicious protection regime by the mid-1960s, as the U.S. presence was regarded as undermining Turkish security and national sovereignty.

BEFORE THE PROTEST CYCLE

The Early Years: The Creation of İncirlik

Although there are still significant gaps in our knowledge about the exact nature of the circumstances under which the U.S. presence in Turkey was created, there is a conventional narrative. In the aftermath of World War II, as the enfeebled British Empire was withdrawing from the Middle East and cutting off aid to Greece and Turkey, the Soviet Union was encroaching on the young Turkish Republic, demanding bases along the Bosphorus and joint control of the Dardanelles, rekindling ancient fears of Russian expansion. The United States stepped into this power vacuum, offering protection from the Soviets, military and economic aid, and membership in NATO, and opening the way to the West. All that Washington asked in return was access to Turkish territory. Within the framework developed by Tilly, the United States was offering legitimate protection against an outside threat and at an unbeatable price. It was a deal that no one else could offer.

Although there are elements of truth to this story, the reality was somewhat different. After the Bolshevik revolution in 1917, Lenin had renounced any territorial claims on the Ottoman Empire and had even supplied Kemal Atatürk with weapons during the War of Independence between 1919 and 1923.[1] Because the Western powers were hostile to both the Soviet Union and the Ottoman Empire, the two became "natural allies" and maintained friendly diplomatic relations from 1917 until at least 1938, even signing a Treaty of Friendship and Nonaggression in 1925.[2]

The United States began establishing facilities in Turkey in 1943 while the Soviet Union was still preoccupied with fighting the Wehrmacht and before it had made any demands regarding the Turkish straits.[3] The first operations were set up in Istanbul by the Office of Strategic Services, the predecessor of the CIA. A second base was established a few months later in a more remote region of southern Turkey, not far from the border of Syria and seven miles from the city of Adana. Neither the Turkish authorities nor the British were informed of this and apparently certain aspects were not even made known to the U.S. Ambassador Steinhardt, although the operation was set up in a house that was ostensibly his summer home.[4] This secrecy was in part attributed to the fact that the station was to collect information on resistance groups in the Balkans without being discovered by the German Consulate next door, but also because the United States was competing with other Allied powers who also wanted to establish a "communications foothold" in Turkey. This meant that many of the agreements with the Turkish authorities were made informally, sometimes even over the telephone.[5] It was not long before the secret "Adana Station" would grow into the now relatively high-profile İncirlik Air Base.

---

[1] The weapons were delivered by Leon Trotsky as commissar for war. While Trotsky was in exile in Istanbul in 1929, he is reported to have said to a visitor: "When Turkey was fighting Greece in the war I helped Kemal Pasha with the Red Army. Fellow soldiers don't forget such things. That was why Kemal Pasha didn't lock me up in spite of pressure from Stalin." Cited in Ahmad 2004: 17.
[2] Ahmad 2004: 16.
[3] Cossaboom and Leiser 1998.
[4] According to Report 57, AACS cited in Cossaboom and Leiser: 1998.
[5] Interview with Cagri Erhan in Istanbul on July 16, 2005.

Although a detailed discussion of Turkey's role in World War II is beyond the scope of this chapter, it should be noted that under the leadership of Atatürk's successor, İsmet İnönü, Turkey remained neutral until February 23, 1945, when it officially declared war on the Axis powers.[6] The Soviet Union was wary of this neutrality and when the Treaty of Friendship was due to expire in 1945, it placed conditions on its renewal. Ankara concluded that it was no longer in a position to continue its tradition of cautious foreign policy and that it would have to align itself with one of the power blocs in formation. It was under these circumstances that relations between the United States and its new-found client gradually became formalized.

On April 5, 1946, before any official aid agreement was signed between the two countries, the USS Missouri sailed down the Bosphorus and anchored in the harbor at Istanbul after traveling across the Atlantic. Although the official reason for the visit was to return the remains of Ahmet Münir Ertegün, the Turkish Ambassador to Washington who had died during the war, the unofficial symbolism of the huge U.S. warship anchored in Istanbul was perhaps more important. The fact that the Turkish government ordered the production of stamps and cigarettes in honor of the USS Missouri and that it was later open for public visitation, is often cited as an indication that the presence of the battleship in the Istanbul harbor was welcomed by the Turkish population as a sign of benevolent protection.[7]

The visit of the USS Missouri was significant for other reasons as well, as students of Cold War history usually cite the visit as one of the first forms of gunboat diplomacy that the United States engaged in after World War II. Although the young Turkish Republic was uneasy about its powerful neighbor to the East, the U.S. Department of the Navy did not necessarily share this threat perception. According to a 1946 report titled "Soviet Capabilities and Possible Outcomes": "[T]he Red Fleet is incapable of any important offensive or amphibious operations ... techniques, tactics, and equipment are far below the Anglo-US standard."[8] According to some, what was perhaps more worrying for the navy than Soviet expansion was that postwar demobilization was not only shrinking the naval forces in Europe (from a peak of 122,900 men in May 1944 to 17,370 in September 1945) but was also threatening to subordinate the navy to the army and air force.[9] At the time, the War Department was pushing for a unification program which would turn the navy into an auxiliary service, and the air force was even claiming that air power made navies obsolete. Secretary of the Navy James Forrestal had other plans. Whereas the army was confined to land and the air force was dependent on access to landing facilities in foreign countries, Forrestal wanted to demonstrate that the

---

[6] For a discussion of Turkey's role in World War II, see Deringil 1989 and Hakki 2007.

[7] Criss 2002.

[8] Cited in Alvarez 1974: 232.

[9] Ibid.

navy could "roam the oceans, protecting vital sea lanes and carrying conflict to the enemy."[10] Although Forrestal could not prevent the merging of the War Department and the Navy Department into the Department of Defense in 1949, with the visit to Istanbul, the navy was able to demonstrate its power as a diplomatic tool and interest in the Mediterranean region.[11]

Several months after the visit of the U.S. warship, Moscow sent a note to Ankara on August 7, 1946, proposing a revision of the Montreux Convention and joint Russo-Turkish control of the straits. According to Melvyn Leffler, Foreign Minister Hasan Saka "breathed a sigh of relief" because there was no demand for bases on Turkish soil. The United States, however, chose to interpret things differently. About a week later, on August 15, U.S. military planners had completed a study code-named GRIDDLE which advocated "every practicable measure ... to permit the utilization of Turkey as a base for Allied operations in the event of war with the USSR." According to the report, "[t]he Turks could slow down a Soviet advance toward Cairo-Suez, thereby affording time for the United States to inaugurate the strategic offensive."[12] Truman was therefore under pressure to extend Marshall aid to both Greece and Turkey. However, not everyone in the administration agreed with the need to establish a U.S. presence in Turkey. George Kennan in the State Department argued that Turkey had not been ravaged by World War II and therefore did not require aid for rebuilding, nor did it face an internal communist movement, nor was it really threatened from Moscow. Even in his memoirs, Kennan continued to insist that the Soviet threat at the time was "primarily a political one and not a threat of military attack" and that the situation in Greece, which was wracked by civil war, in no way pertained to Turkey.

On February 24, 1947, the British Ambassador Lord Inverchapel officially informed U.S. Secretary of State George Marshall that Britain could no longer provide aid to Greece and Turkey. The governments in Ankara and Athens were not to be informed until Washington had decided whether it would provide aid and hence take on the mantle that the British were leaving behind. Within three days, President Truman had decided that aid would be forthcoming.[13]

In a speech before Congress on March 12, 1947, President Truman requested $400 million in aid for Greece and Turkey, as part of what would become known as the Truman Doctrine.[14] The president had been convinced by the

---

[10] Alvarez 1974: 230.

[11] In 1949, the army, navy, and air force were consolidated into the Department of Defense and placed under the leadership of the Secretary of Defense.

[12] The archival source is cited in Leffler 1992: 551.

[13] McGhee 1990: 19ff.

[14] In his speech before Congress, Truman declared: "I believe that it must be the policy of the United States to support free people who are resisting attempted subjugation by armed minorities or by external pressures." The Truman Doctrine was the beginning of the U.S. commitment to defend anti-communist regimes. Zürcher interprets the change from a one-party to a multiparty regime in Turkey as a result of both domestic and U.S. pressure. Zürcher 2003: 218ff.

Pentagon's arguments, and not the admonitions of George Kennan, the architect of the containment policy. Leffler interprets the U.S. policies at the time as
follows: "American fears did not stem from aggressive Soviet moves against
Turkey. The Soviets had done little more than send a diplomatic note. The real
problem was that there loomed gaping vacuums of power in this part of the
world resulting from the decline of British power."[15]

Approximately two years later, in April 1949, the North Atlantic Treaty
Organization was founded in Washington, DC. The original twelve members
were the United States, Belgium, Britain, Canada, Denmark, France, Iceland,
Italy, Luxemburg, the Netherlands, Norway, and Portugal. Membership was
not extended to Greece and Turkey. Less than a week later, Foreign Minister
Necmettin Sadak complained to Dean Acheson that the U.S. bases in Turkey
increased the likelihood of a Soviet attack, and yet there was no guarantee that
the United States or any of the other NATO members would defend Turkey
as long as it was excluded from the alliance. Already at this very early stage,
the U.S. bases in Turkey were seen as a potential liability or even as potential
targets. The dual nature of foreign bases – that both provide and potentially
undermine security – was recognized from the beginning.

The 1950s: The Expansion of the U.S. Presence

The first aid agreement between the two countries was signed in July 1947 as
part of the Marshall Plan. In October of 1949, the Joint American Military
Mission for Aid to Turkey (JAMMAT) was established.[16] A major expansion of
the U.S. presence occurred during the following decade, the 1950s. In the first
truly independent multiparty elections held in May 1950, the Democrat Party
(DP) under the leadership of Adnan Menderes deposed the Republican People's
Party (CHP) which had ruled the country since its founding in 1923. In July
1950, the newly elected Prime Minister unilaterally decided to send 5,000 combat troops to Korea, demonstrating that Turkey was willing to defend Western
interests, although NATO had made no such commitment to Turkey.

Soon afterward, in the spring of 1951, the U.S. company Metcalfe, Hamilton,
and Grove began construction of the İncirlik base by clearing a large plot of
land on the outskirts of Adana which had been cultivated as a fig orchard.[17]
Here it is important to emphasize that this was still before Turkey became a
member of NATO in 1952, because often the history of the U.S.-Turkish military partnership is subsumed under the larger story of the development of
NATO. For the purposes of this chapter, however, it is important to emphasize
that the foreign military presence in Turkey is essentially an American phenomenon. Some observers have even claimed that Turkey's NATO membership

---

[15] Leffler 1992: 124.
[16] Criss 1993: 342.
[17] İncirlik in Turkish literally means "place of the fig orchard."

is primarily an "external packaging" of what is essentially a bilateral defense relationship with the United States.[18]

In February 1951, George McGhee, who was soon to become ambassador, visited Turkey. President Celal Bayar indicated to him that if Turkey were not admitted into NATO, it might consider returning to its prewar policy of neutrality. Although military planners would have preferred keeping the advantages of the U.S. presence in Turkey without the obligations of an alliance, the prospect of Turkish neutrality was enough to make them change their minds.[19]

The Korean War was the first time that communist and Western forces engaged in armed combat. The government in Ankara had apparently decided that if war could break out in Asia, perhaps it could also happen elsewhere, and that it needed a security guarantee if it was going to continue to allow the United States to increase its military presence. Whether or not the Soviet Union ever seriously considered launching an attack against Turkey is less important for the current purpose than to establish the fact that it was perceived as a threat. The threat was considered so real that the Turkish government insisted on being included in the NATO alliance as a full member.

The entrance of Turkey into NATO would not have happened without support from Washington. The Europeans, who were initially unenthusiastic about including a southern Mediterranean country with a Muslim-majority population into NATO, were convinced by the U.S. argument that Turkey's eighteen divisions could divert Soviet forces to southern Europe. Furthermore, Turkish troops were inexpensive to maintain.[20] The invitation to join NATO was approved by the Turkish parliament by a vote of 404 to 0, with 1 abstention.[21]

Attempts to establish U.S. bases elsewhere in the Middle East were less successful. For example, when Secretary of State John Foster Dulles attempted to convince Gamal Abdel Nasser that Egypt should join an alliance with other Middle Eastern states to counter the Soviet Union, Nasser allegedly replied: "The Soviet Union is more than 1,000 miles away and we've never had any trouble with them. They have never attacked us. They have never occupied our territory. They never had a base here. But the British have been here for 70 years."[22] Nasser argued that he would not be taken seriously, indeed even become the laughing stock of the country, if he tried to convince the Egyptian people that the USSR was a greater threat than the British.

Compared to other NATO members in Western Europe, Turkey was an underdeveloped country characterized by a small Westernizing elite and a

---

[18] Supplement on Turkey, *Financial Times*, May 14, 1984.
[19] Ahmad 2004: 32.
[20] In 1986, the cost of outfitting a Turkish soldier was seven times less expensive than for an American G.I. Karasapan 1989: 5.
[21] Vali 1971: 126.
[22] Cited in Brands 1993: 53.

large rural population. However, it was believed to be "on the right track" in terms of the process of modernization. Although Daniel Lerner is now often dismissed as a Cold War propagandist, his research is important to mention in this context because it contributed to this assessment of Turkey. Lerner began his best-known book *The Passing of Traditional Society: Modernizing the Middle East* with a study of Balgat, a village eight kilometers outside of Ankara.[23] Lerner contrasts the grocer (*bakkal*) with the village elder or chief (*muhtar*). In 1950, the grocer appeared to be the only person in the small village who seemed to possess the traits of a "modernizer," whereas the chief was clinging to the traditions of the past. Between the time of the original research trip in 1950 and the second trip in 1954, Balgat had undergone a period of rapid development. To categorize them as either "modernizers," "transitionalists," or "traditionalists," villagers were asked to answer questions about their attitudes toward Americans and other foreigners, whether they had ever listened to Voice of America, BBC, or other foreign broadcasts, and whether they preferred American or Turkish movies. By 1954, Lerner concluded that the traditional way of life was "passing." In 1961, a few years after the study was completed, the United States began construction of an air base in Balgat.

The growth of the U.S. presence was accompanied by large economic and military aid packages. Under U.S. supervision, the Ministry of Defense was reorganized, the War Staff College in Istanbul resumed full-time instruction with American, British, and Turkish faculty, and many Turkish officers were sent to the United States for further training. Perhaps more significantly, the Turkish armed forces underwent a full-scale modernization program. When U.S. military advisers first arrived in Turkey, they found airfields with landing strips made with low-grade asphalt that could not support the heavier U.S. aircraft and which were not equipped with runway lighting or control towers. The military also reflected the huge class divisions within Turkish society: many of the conscripts were from peasant backgrounds, illiterate and unaccustomed to handling mechanized equipment whereas the higher-ranking officers were highly educated and often spoke at least one foreign language. As opposed to the U.S. military, where promotions allow at least some degree of upward mobility, the enormous gulf between the Turkish rank-and-file and their commanders meant that even talented recruits had little chance of ever joining the officer corps. A full-scale description of the modernization of the Turkish armed forces would extend the scope of this chapter, but Craig Livingston has summed up the technological impact of the military assistance program. According to Livingston, the United States helped transform the Turkish Air Force "from an antiquated, inefficient, almost harmless collection of airplanes to an air arm boasting state-of-the-art propeller driven aircraft, jets, a solid

---

[23] See Lerner 1958. For a critical discussion of the connections between Lerner's development communication research and his propaganda research, see Bah 2008.

logistical base, and a key place within the alliance."[24] This was of course only possible with large financial outlays: from 1950–1974, U.S. military aid to Turkey averaged $165 million per year.[25]

In addition to military aid, the United States also supervised other development programs. By most accounts the highway program was by far the most successful, as it connected the entire country together, creating the infrastructure for further economic development. Because Prime Minister Adnan Menderes was elected in part because he claimed to represent the Anatolian majority (as opposed to the Western elites), the project of connecting the villages with the urban metropoles was politically very important. Some observers have claimed that aid from Washington was important, perhaps even essential, in consolidating Menderes' rule, unpopular as he was among the traditional Kemalist power-holders. During the 1954 elections, his opponents argued that U.S. aid was protecting Menderes from his own mistakes and the phrase "If Allah does not provide, America will" became a common refrain in opposition circles.[26] Nevertheless, Menderes was reelected twice, in 1954 and 1957, and remained Prime Minister for ten years, until he was deposed by the 1960 military coup.

During his ten-year-long incumbency, the U.S. presence expanded. Because of the secrecy surrounding military installations and because of the fact that the U.S. presence changed over time, it is very difficult to find a definitive list of all of the facilities that existed at any certain point in time. By compiling information from various sources, however, it is possible to put together the following picture of the U.S. presence during the Cold War (Map 2.1).

In addition to the air base in İncirlik, other bases were built in Istanbul, Ankara, Balgat, Bandirma, İzmir, Karamürsel, Çiğli, Antalya, and Diyarbakır. Radar stations were constructed along the Black Sea at Sinop, Samsun, and Trabzon, as well as in the interior in Belbasi and Diyarbakır. Naval facilities and storage centers were built at İskenderun (Alexandretta), Marmaris, and Yumurtalik in the Mediterranean, İzmir in the Aegean, and Haydar Paşa and Çanakkale in the straights. Communication nodes were also constructed in Alemdağ, Balıkeşir, Elma Daği, Karataş, Kargaburun, Kurecik, Mahmurdağ, Pirinclik, Sahin Tepesi, and Yamanlar. Other facilities were built in Çakmaklı, Eskişehir, Golçuk, İzmit, Manzaralı, Murted, and Ortakoy. The capital city of Ankara played a major role as the headquarters of the U.S. Logistics Group, Turkey (TUSLOG). William Hale described these installations as "US-cum-NATO facilities." Finally, the U.S. Air Force stationed strike aircraft armed with tactical nuclear weapons on Turkish soil. By the late 1960s, there were almost 30,000 U.S. military personnel on Turkish territory.[27] For a country which was proud that it had never fallen under the colonial yoke, this was no small feat.

[24] Livingston 1994: 778.
[25] Duke 1989: 274.
[26] Harris 1972: 81.
[27] Data compiled from Duke 1989: 282ff; Hale 2000: 123; and "Military Bases of the USA and FRG in Europe," *International Affairs*, December 1961: 108–110.

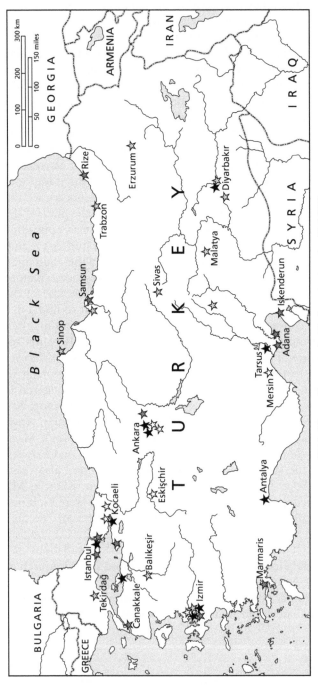

MAP 2.1. US Military Presence in Turkey during the Cold War.

In addition to the land-based forces, the Sixth Fleet roamed the Mediterranean as a highly mobile military base featuring two aircraft carriers. The carrier-based aircraft were capable of dropping conventional or nuclear weapons on any target in the sea and more than 1,000 miles from the carriers, which brought areas in the southern Soviet Union, Romania, Bulgaria, and Hungary within reach.

In sum, the U.S. presence in Turkey was composed of army, air force, and navy units as well as intelligence-gathering facilities. The U.S. presence featured installations that were highly visible as well as others that were less visible, and some that were entirely secret. The installations along the Black Sea had a low profile, and yet they were extremely important as they monitored Soviet rocket launchings at the main missile sites near the Aral Sea. A report from the mid-1960s described the system as being "so effective that no Soviet rocket could get off the ground without its being known within a few minutes at the North American Defense Command in Colorado Springs and at the CIA and the White House."[28]

Some of the more secretive aspects of the U.S. presence included CIA operations and the establishment of a special warfare department. Daniele Ganser's book *NATO's Secret Armies: Operation Gladio and Terrorism in Western Europe* represents one of the first thorough investigations of the so-called stay behind armies that were created throughout Europe during the Cold War. Their purpose was two-fold: to create a type of backup army to support the official NATO forces in the event of a Soviet invasion and also to counter internal enemies. In Turkey, the United States enlisted the help of ultra-nationalist Colonel Alparsan Türkeş. In the early 1950s, Türkeş had already set up the Şeferberlik Taktik Kurulu or Tactical Mobilization Group which was located in the building of the American Aid Delegation within the JUSMATT complex in Ankara. In 1965, the organization was restructured and renamed Special Warfare Department (*Özel Harp Dairesi*). It has been considered as forming one of the key institutions of the so-called deep state (*derin devlet*). According to his own testimony, Prime Minister Ecevit only became aware of the organization in 1974 when military officials asked for funding after a decrease in U.S. aid. Subsequent attempts by Prime Minister Ecevit and others to uncover the extent of their activities were fruitless. It was not until the Gladio revelations of 1990 that the organization was exposed, after which it was renamed again as the Special Forces Command (*Özel Kuvvetler Komutanlığı*).[29] As I have argued in Chapter 1, the relationship between the U.S. presence and the non-U.S. citizens who live under its security umbrella is inherently contradictory. As this example illustrates, the U.S. military often operated outside the realm of accountability, disenfranchising not only ordinary citizens, but even elites in countries that hosted U.S. bases. The protectariat therefore includes both elites and non-elites who are defined by their relationship to the U.S. protection regime, or the means of violence, as opposed to the means of production.

---

[28] Cited in Lewis 1976: 31.
[29] Ganser 2005: 226.

The exact nature of the activities of this organization are still being researched and are therefore still disputed. Based on training manuals, however, it is possible to delineate their intentions. One such manual was written by David Gallula, a Tunisian-born French military officer who was influential in developing the theory of counterinsurgency warfare, based on his experiences in Indochina, Greece, and Algeria. "The Repression of Popular Uprisings: Theory and Praxis" was translated into Turkish and distributed within the military in 1965.

> Our security isn't threatened just by external attacks. In addition there are other threats which are much more dangerous ... these camouflaged attacks ... are at times civil wars, at times uprisings, but also democratic and reformist movements. It is our intention to prevent the rise of these movements.[30]

The bulk of the U.S. military presence, however, was not clandestine but existed in the form of large buildings, rumbling military vehicles, and support facilities including schools, shopping malls, bowling alleys, and sport facilities concentrated in urban areas. Such an obtrusive foreign military presence necessitated a legal framework. One of the treaties that regulated the precise nature of the bases as well as the responsibilities and privileges of the U.S. personnel was known as the Military Facilities Agreement and was signed in 1954. The full text of the treaty was never disclosed to the public. According to Gonlubol, one of the various reasons the treaty was to remain secret was that if it had become publicized, it may have reminded the Turkish people of the capitulations granted to foreigners at the end of the Ottoman Empire.[31] Another aspect of the treaty that was not disclosed was that the Turkish government not only made 35.9 million square meters of land available to the United States but also assumed the financial burden of the expropriation costs of the land, the responsibility for the protection of the bases, and the maintenance of environmental security, which was estimated to be about $11 million annually.[32]

Despite the fact that the most sensitive aspects of the treaty were not known to the public, there were already murmurs of discontent within the populace. Although the Korean War is usually held up as a shining example of Turkish-U.S. military cooperation, some had questioned the prudence of Turkey's involvement in a war in East Asia. The Turkish Brigade suffered a large number of casualties and the phrase "*Kore yolunda öldü*" ("He died on the way to Korea") came to mean "He died for nothing" in everyday usage.[33] However, these popular attitudes did not result in any large-scale organized anti-war activities; only a small group called "*Barişseverler Cemiyeti*" (Association of Turkish Peace Lovers) issued a flyer against the Korean War and were criticized by Prime Minister

---

[30] Cited in Karasapan 1989: 7.
[31] Capitulations were extraterritorial privileges that allowed foreigners to trade freely, practice their religion, and live by their national laws without paying taxes and without regard to the central Ottoman authority.
[32] Gonlubol 1975.
[33] Interview with Çağri Erhan in Istanbul on July 16, 2005.

Adnan Menderes himself and imprisoned.[34] Such limited anti-war demonstrations, however, were a far cry from the anti-U.S. rumblings elsewhere in the region. In May 1953, Secretary of State Dulles made a trip to the Middle East where the official welcome mat that was rolled out for him was marred by popular unrest: in Cairo demonstrators threw eggs at him, in Beirut the Lebanese government asked him not to visit the American University because they feared Communist demonstrations, and he skipped a visit to Iran entirely, perhaps because plots to overthrow Prime Minister Mossadegh were already underway.[35] By comparison, he considered Turkey "a highly important American ally" and even said that "no country, people or government in the world today reacted more firmly and effectively to the Soviet Communist menace than did the Turks."[36]

These were not merely diplomatic pleasantries, but reflected the fact that Turkey was a willing participant in many U.S.-led plans for the Middle East, such as the Baghdad Pact, which later became known as the Central Treaty Organization (CENTO) after Iraq withdrew following the overthrow of the monarchy in 1958. For the Pentagon, Turkey quite literally formed a bridge between East and West, linking NATO to Iran and Pakistan, and giving geographical shape to the military alliances "containing" the Soviet Union. Turkey was expected to defend the Transcaucasian border with the Soviet provinces in the East (the longest border with the USSR of any NATO country), to cooperate with its long-time rival Greece against a possible aggression coming from the direction of Bulgaria in the West, and to guard passage through the straights. Last but not least, Turkey was used as a launch pad for interventions in the Middle East. In 1958, Ankara allowed the İncirlik Air Base to be used as a staging point for the U.S. intervention in Lebanon, when the United States landed 16,000 marines in Beirut, although this operation clearly exceeded the parameters for using Turkish facilities established by NATO.[37] Around the same time, other countries in the region – such as Egypt and Syria – were drawing closer to the Soviet Union. By 1956, the last British soldiers had withdrawn from Port Said, their outpost along the Suez Canal. The British Empire had thereby lost its largest overseas base in the world. Also around the same time, Cairo received its first arms shipment from the Soviet Union. Other countries in the region also sought assistance in their struggle against the remaining British colonial presence in the region. The Menderes government, acting in concert with Washington, moved troops to the border of Syria to threaten the nationalist regime in Damascus, prompting the USSR to mobilize troops in the border

---

[34] *Hürriyet*, July 29, 1950.

[35] In March 1953, Secretary of State Dulles gave directions to his younger brother Allen Dulles, who headed the CIA, to begin attempts to oust the democratically elected Prime Minister Mossadegh. The initial decision to remove Mossadegh came from the British, who were upset at their loss of control over the Iranian oil industry.

[36] Foreign Relations of the United States (FRUS) 1958–1960, Vol. X, Part 2: 739ff.

[37] In his memoirs, President Eisenhower recalled how Menderes was even willing to send Turkish troops to counter the Iraqi coup in 1958. Cited in Evriviades 1998: 38ff.

region near Turkey.[38] Compared to much of the Arab Middle East, which was increasingly coming under the influence of Nasser's pan-Arab nationalism, Turkey was indeed an important asset.

However, not everyone in Turkey was enthusiastic about becoming a junior partner of the United States. In 1958, the Operations Coordinating Board wrote a report called "Operations Plan for Turkey" (NSC 5708/2). The report noted a disparity between the pro-U.S. policies of the government and the increased tensions from below. According to the report, although the Turkish government gave approval for use of the İncirlik base during the Lebanese operation,

at the same time there are indications that local attitudes toward U.S. service personnel are changing from acceptance to hostility. Although incidents involving U.S. service personnel provide the focal point for outbursts of popular resentment and bitter press comment, wide disparity in standards of living, sharply contrasting social mores, a formidable language barrier and alleged special privileges for American military personnel are continuing factors exacerbating community relations.[39]

The same report underlined that it was necessary to take certain steps to ensure that the problems would not get out of hand. Observing that there were already 13,000 U.S. personnel stationed in Turkey, the report suggested that every U.S. agency should make sure that the number of U.S. employees in Turkey "does not exceed the minimum required to achieve US objectives." Furthermore, it was necessary to "strengthen information activities aimed at maintaining Turkish public acceptance of U.S. military personnel and installations in Turkey."[40] Finally, the report also encouraged the continuation of U.S. support for the development of modern labor-management relations in Turkey. In particular, this meant supporting ties between Turkish labor and anti-communist international labor federations such as International Confederation of Free Trade Unions (ICFTU) and International Trade Secretariats (ITS) so that the goals of Turkish labor unions would be "economic rather than political."[41]

Five months after the report was published, Secretary of State Dillon wrote a letter to the Secretary of Defense McElroy in which he begins by describing how the cooperation between the United States and Turkey has led to an increase in the U.S. presence there. The Secretary of State then continues to address the problems surrounding the U.S. presence in Turkey:

It is inevitable that the presence in any foreign country of so many Americans would involve difficult problems of community relations. In Turkey, however, the increasing number of incidents has caused me, and I am sure has caused you, very real concern, as has the disquieting rise in antagonism on the part of the Turkish public toward foreigners on their soil which has resulted from these incidents. While we have had, and continue to have, excellent cooperation on the part of the Turkish authorities, I am

---

[38] Ahmad 1977: 397.
[39] Foreign Relations of the United States (FRUS) 1958–1960, Eastern Europe; Finland; Greece; Turkey, Vol. X, Part 2: 768.
[40] Ibid: 782.
[41] Ibid: 776.

greatly concerned that if the situation continues not only might general Turco-American relations be impaired but, with particular reference to our military operations in that country, we might find ourselves in real difficulty in maintaining highly important facilities which we now enjoy.[42]

To prevent a loss of U.S. facilities, Dillon suggested reducing the number of U.S. military personnel in Turkey, only dispatching other personnel for "essential purposes," and creating an interagency working group including the three service arms (the Central Intelligence Agency, the United States Information Agency, and the International Communications Agency) as well as the Defense and State Departments to monitor the situation and find solutions.[43] Although these documents demonstrate the U.S. preoccupation with its own problems, these events also contributed to the crisis of the Menderes regime.

The 1960 Coup

The more general issue of Turkey's alliance with the United States became highly politicized toward the end of the decade as opposition to the Democrat Party increased. In addition to charges of corruption and heavy-handedness toward political dissent, the CHP accused the DP of "selling out" the nation to the Americans. Prime Minister Menderes had certainly benefited from the large U.S. aid programs that had flowed into Turkey, helping to consolidate the DP's ten-year rule as the first political party to oust the CHP from power. In return for economic and military aid, Menderes and his administration gave the United States free rein to establish whatever facilities it deemed necessary, leading some to argue that basing access was the most fundamental aspect of the bargain struck between Washington and Ankara.[44]

In 1957, a group of conspirators was discovered within the army. And in 1959 Menderes signed a Bilateral Agreement with Washington that guaranteed U.S. intervention in the event that communists or other forms of "internal aggression" threatened the integrity of the regime.[45] For this reason, Menderes may have believed that he would prevail despite the political wrangling. The CIA, although apparently concerned about the situation, seems to have shared his assessment. In a letter from the Director of the CIA Allen Dulles to President Eisenhower, Dulles described how the leader of the CHP, İsmet İnönü, a former army general, principal aide to Ataturk in founding the republic, and Ataturk's political

---

[42] Foreign Relations of the United States (FRUS) 1958–1960, Eastern Europe; Finland; Greece; Turkey, Vol. X Part 2: 802.

[43] The Department of Defense responded to the Secretary of State's letter by insisting that the number of personnel could not be reduced, that the number of incidents between U.S. personnel and locals was not as high as Dillon had argued and therefore did not agree to an interagency committee, but only to an informal State-Defense working group.

[44] Pope 1997: 84–93.

[45] Ahmad 1977: 297; Harris 1972: 221–223.

heir, believed that Turkey was not getting the quid pro quo that it deserved from the United States. Although U.S. officials were wary of the delicate political situation, a report prepared by the CIA, as well as the intelligence organizations of the Departments of State, the Army, the Navy, the Air Force, and the Joint Staff, came to the following conclusion: "Barring drastic economic deterioration or extreme political provocations, the chance of a military coup remains slight."[46] Fifteen months later, on May 27, 1960, the Turkish military removed Menderes from power and created the Committee of National Unity (CNU).

Some in the opposition feared that the United States may intervene to prop up Menderes, but this did not happen. For Washington, the Menderes government was only important as long as it could safeguard its interests. In fact, the U.S. officials displayed a large degree of flexibility in this regard. The U.S. ambassador to Turkey, Fletcher Warren, wrote a letter to the Assistant Secretary of State for Near Eastern and South Asian Affairs in August 1960 in which this became quite evident:

We were as helpful to the Turkish people, through the Menderes Government as we were able to be. Now we can say, by a stretch of the imagination, that the Turkish people has given us the Gursel Government with which to work. We intend to work with it just as loyally and faithfully as we did with the Menderes Regime.... It might be said in parentheses that, when the Gursel Government goes, we shall endeavor to be in a position to work in the same friendly, cooperative way with the succeeding government (if it is not Commie).[47]

After the provisional government took over the reins from Menderes, the United States became worried about the political stance of the CNU, although a declaration had been made immediately after the coup that Turkey would remain loyal to NATO and its foreign alliances. Despite the fact that all thirty-eight members of the CNU had attended army schools in the United States, U.S. officials had not been able to establish close relations with the new regime.[48] In the same letter, U.S. Ambassador Warren complained that he and other diplomats did not even know who the key figures were within the CNU. "We were not sure whether Gursel was a stooge, a disgruntled military man, or a patriotic Turk whom unusual forces had brought to the head of the Government." Warren also expressed his concern over the compulsory retirement of the higher-ranking officers within the Turkish military: 90 percent of the generals, 55 percent of the colonels, 40 percent of the lieutenant colonels, and 5 percent of the majors were being sent home. Although the official reason was to transform a top-heavy military, the officers who were being retired were also the same people whose loyalty to the provisional government may have been in question.[49] U.S. officials seemed to be less concerned about the

[46] Foreign Relations of the United States (FRUS) 1958–1960, Vol. X, Part 2: 785.
[47] FRUS 1958–1960, Vol. X, Part 2: 878.
[48] FRUS: 853.
[49] FRUS: 869.

political implications of the purge, than they were about the size of the officer corps. At the time the Turkish Armed Forces only had 60 percent of authorized officer strength, and cutting 10 to 15 percent would weaken the military force that was already deemed inadequate. Despite these concerns, the United States released financial aid to cover the pensions of the retired officers.

The implications of the coup in terms of U.S. access to facilities in Turkey were summarized in a National Security Council Report (NSC 6015/1) from October 1960: the report warned that the new regime may be "less inclined" to informal basing agreements and may "look more closely at U.S. use of Turkish military facilities."[50] Under these circumstances, the NSC listed five U.S. objectives for Turkey in the following order: (1) preserve Turkey's territorial integrity, (2) secure continued access to military facilities and Turkey's cooperation within NATO and CENTO, (3) maintain the Turkish Armed Forces, (4) achieve a democratic government, and (5) pursue economic growth that will eventually eliminate Turkey's need for economic aid.[51] In short, maintaining access to military bases in Turkey was given priority over fostering economic growth or democratization.

According to conventional wisdom, the end of World War II brought prestige to the United States, whereas its allies felt more or less secure under the protective umbrella of NATO and U.S.-Turkish relations were characterized by warmth and sympathy on both sides. The archives, however, would tell a different story. The decade of the 1950s in many ways did not auger well for the continuing U.S. presence and foreshadowed problems in the 1960s. Although the Korean War was unpopular among a small minority, the discontent caused by the U.S. presence in Turkey was more widespread and much more troubling for U.S. authorities, as no one less than the Secretary of State saw it necessary to suggest policy changes. Therefore, maintaining access to the military installations was given precedence over fostering economic growth or democratization, as the NSC report indicates. Finally, even in the absence of social movements, the U.S. authorities were apparently already concerned about popular attitudes toward the United States just a few years after the İncirlik base had opened, and long before the tribulations of the 1960s and 1970s.

THE OPPOSITION CYCLE BEGINS

1960s: The Beginnings of Organized Protest

The 1960s were marked by a number of international incidents which drew attention to Turkey in general and İncirlik in particular. Perhaps the most conspicuous of these was the U-2 spy plane incident in 1960 because it precipitated an international crisis involving not only the United States and the Soviet

---

[50] FRUS: 889.
[51] FRUS: 896.

Union, but also Norway, Pakistan, and Turkey. On May 1, 1960, Francis Gary Powers, who had been stationed at İncirlik for more than four years, took off from Peshawar in Pakistan in a U-2 reconnaissance aircraft with the intent of collecting intelligence in Soviet airspace. However, rather than landing in Bodo Norway as planned, the plane was allegedly shot down well inside Soviet territory and the pilot was subsequently tried and sentenced to imprisonment in the USSR.[52] On May 13, Khrushchev issued the following warning:

> These countries that have bases on their territories should note most carefully the following: if they allow others to fly from their bases to our territory we shall hit at these bases … if you lease your territories to others and are not the masters of your land, of your country, hence we shall have to understand it in our way.[53]

The governments of Turkey, Norway, and Pakistan all claimed they had been uninformed about the fact that the U.S. military was carrying out this type of intelligence gathering. Each country subsequently declared that "every effort would be made to prevent the use of bases in their respective territories for unlawful reconnaissance flights over the Soviet Union."[54]

After the U-2 event, which demonstrated among other things that the U.S. military authorities were playing according to their own rules rather than those laid out in the Status of Forces Agreement, the Cuban Missile Crisis in 1962 and the Johnson letter of 1964 further undermined Turkish confidence in the United States. Although the Turkish generals had no illusions about the fact that they could not exercise the same amount of freedom as their American counterparts, the High Command was taken aback when President Johnson warned them not to intervene to protect the Turkish minority in Cyprus after the outbreak of intercommunal violence, although their right to do so was recognized in the Guarantee Treaty of 1959/60.[55] Because Johnson stressed that NATO was not necessarily obliged to protect Turkey if an intervention prompted Soviet interference, Ankara began to question the purpose of NATO if it did not provide security in return for the use of Turkish territory.[56]

---

[52] This version of the story is not undisputed. Because it is unlikely that a pilot could be shot down at a height of 80,000 feet and then live to tell about it, James A. Nathan claims that Powers intentionally landed inside Soviet territory and that the U-2 incident was planned to undermine the emerging détente between the United States and the Soviet Union. By having an American pilot violate Soviet air space and then explain to the public that he had been doing so on a regular basis for almost five years as part of his overseas assignment, the groundwork was laid for a heightening of Cold War tensions and a perpetuation of the arms race. Whether the event was planned by Eisenhower himself or Pentagon officials is unclear; however, less than one year later, President Eisenhower gave his farewell speech in which he warned about the growth of the "military industrial complex." Nathan 1975.

[53] Reported by the Soviet News Agency, TASS, and cited in Woodliffe 1992.

[54] Cited in Woodliffe 1992: 265.

[55] This treaty was signed by Turkey, Greece, and the UK at the 1959 London Conference; it went into effect when Cyprus became independent on August 16, 1960.

[56] Particularly veterans of the Korean War, who had fought alongside Americans a mere ten years earlier, felt betrayed by the United States.

Between 1947 and 1965, fifty-four bilateral agreements were signed between the United States and Turkey, forty-one of which were secret.[57] The U-2 spy plane incident and the Cuban Missile Crisis were both significant in this regard in that they both revealed U.S. military secrets. Until Gary Powers was shot down it was not known that the United States had been carrying out such espionage activities from Turkish bases and until Kennedy promised to remove the missiles stationed in Turkey in return for the Soviet Union withdrawing its missiles from Cuba, it was assumed that these missiles were under joint U.S.-Turkish control and served to deter a Soviet attack and therefore increase Turkey's security. The unilateral removal of these missiles seemed to undermine not only Turkey's security, but also its sovereignty. Both the U-2 spy plane incident and the unilateral removal of the missiles underlined the extent to which the U.S. military operated in a realm that was off-limits even for Turkish elites. The American protection regime disenfranchised the very people it claimed to protect, and even increased the possibility of a Soviet attack, as Khrushchev's exhortation made clear.

Many scholars have argued that the so-called Johnson letter, however, was the turning point. In this letter President Johnson warned Prime Minister Inönü not to intervene in Cyprus, hence rebuking previous guarantees of protection as a NATO member. According to the conventional argument, this alleged betrayal was of a magnitude that popular opinion began to shift and the image of "Uncle Sam" was slowly replaced by that of "Imperialist America."[58] Because the British had sovereign bases on Cyprus, which the United States could use, and both Greece and Turkey were members of NATO, there was no threat of losing facilities in Cyprus, hence the United States was in a win-win situation, whereas Turkey was not.

The U-2 event, the Cuban Missile Crisis, and the Johnson letter clarified two things. First, what constituted a threat to Turkey would be determined not by Turkey but by the United States, and hence protection would be offered only against that particular US-defined threat. Second, the type and nature of the protection would also be determined by the United States and could be modified without consultation, as the removal of the missiles demonstrated.

As explained, however, popular attitudes had begun to shift already in the 1950s, long before Cyprus became an issue. By 1966, elite attitudes began to shift as well, as the Turkish government was requesting revisions in twenty of the fifty-four bilateral agreements.[59] Beginning around the same time, such tensions at the state level were upstaged by tensions emanating from below. Social resistance to the U.S. military presence became increasingly organized.

Forms of protest against the U.S. presence ranged from more traditional opposition within the parliament to forms of civil disobedience and violent attacks. I have classified these types of social unrest into four categories: (1)

---

[57] *Turkish Daily News*, October 17, 1966.
[58] Candar 2000: 140.
[59] *Turkish Daily News*, October 17, 1966.

opposition within the parliament (parliamentary opposition), (2) strikes of Turkish workers at American military facilities (labor unrest), (3) protests against the Sixth Fleet (civil disobedience), and (4) attacks against U.S. personnel and installations, including kidnappings (armed resistance/violent attacks). Finally, there were occasional outbursts of rioting or other clashes that I will not explore in depth because they were not part of an organized movement, although they were significant in other ways.

Turning our focus in the other direction, and considering briefly popular attitudes of Americans toward Turkey, further questions the received wisdom about the U.S. presence in Turkey and elsewhere. Although the global U.S. presence is often naturalized as it is referred to as an overseas "posture" or "footprint," it is helpful to remember that the presence of large numbers of U.S. soldiers and sailors in Turkey was by no means natural, and was not experienced as such by either side. Until now, only one single ethnographic study has been done on the U.S. presence in Turkey. It was carried out between 1965 and 1967 and was published in 1969. As part of her dissertation research, Charlotte Wolf studied the U.S. presence in Ankara as what she termed an "ethnic community" and argued that individual interactions with the Turkish community were almost entirely mediated through the structure of the military. By way of introduction, she begins by describing the old part of Ankara as she experienced it, and, although coming at the beginning of what is meant to be an objective analysis, it was apparently intended to be a word of warning to her fellow citizens:

The old section, spreading out from the broad boulevard with its proper buildings, quickly dissolves into labyrinthine disarray and the wild helter-skelter of bazaars, lean-to shops selling shoelaces and combs and cheap assorted items; antique stores, mosques, crumbling houses of tiny flats without electricity or water, and a frenetic traffic of crowds, cars, trucks, and horse-drawn carts.... A walk through this section can become a nightmare of weird and unintelligible exotica, of apprehension, of cultural disorientation.... Here the rawness and austerity of life do not keep their distance, have not been schooled in the Western amenities, but with sickening insistency penetrate all one's senses.... The newness, the ugliness of superimposed modernity is here, yet old Ankara has a way of seeping in along the edges and animating the Westernized version with Turkish flavor. This new area is the center of Turkish pride in what was willfully wrought from the ancient clay of Anatolia; this is the center where the foreigners shop and live; and this is where the American military community members live out their tours of duty, distinct and insulated from much of that which is really Turkish.[60]

This brief excerpt of Wolf's ethnographic study not only belies a certain orientalism, but also illustrates the fact that even the most normal or benign relations at the state level can be accompanied by a jolting cultural disorientation of the citizens of the respective states on the ground. In the following narrative, I hope to illustrate how an accumulation of protest activities at the

---

[60] Wolf 1969: 34–35.

micro-level eventually reverberated back up to the state level and led to substantial changes in the bilateral relations between the United States and the Republic of Turkey.

Opposition within the Parliament – Türkiye İşçi Partisi

The 1960 coup resulted in a new constitution that defined Turkey as a secular, social, and democratic republic. It went into effect in 1961 and has been described as the most democratic constitution the country had ever had. The Türkiye İşçi Partisi (TIP) was established on February 13, 1961, the same day that the Ministry of the Interior lifted the ban on political activities. Although the literal translation would be "Workers' Party of Turkey," TIP is often referred to as the Turkish Labor Party.[61] It was not only the first legal socialist party in Turkey, but was by some accounts also the most significant.[62] On a more general level, it was also one of the first signs of the birth of civil society in a country stifled by one-party rule and bureaucratic stagnation. In August 1962, Mehmet Ali Aybar became chairman of the party.[63] As a legal scholar and practicing lawyer, Aybar held the 1961 constitution in high regard and under his stewardship the Türkiye İşçi Partisi remained dedicated to change through the parliamentary process. The First Party Congress was held February 9–19, 1964 in Izmir and defined the party program as favoring a "non-capitalist path of development" (*kapitalist olmayan yol*). One of the main themes was the nationalization of various branches of the economy.[64]

During the first half of the 1960s, the goals of TIP have been interpreted as being part of the general so-called social awakening that was sweeping Turkey.[65] Among other things, the goals of this "awakening" included increasing popular participation in economic and social life and expanding production through state planning. One important achievement in this regard was gaining the right to strike and collective bargaining, which was secured through the efforts of Türk İş and Bülent Ecevit, then the Republican Minister of Labor.[66]

---

[61] Feroz Ahmad makes the point that very few leftist institutions in Turkey referred to themselves as "Turkish" but rather as being "of Turkey," whereas placing an emphasis on "Turkishness" was more likely a signifier of the right. As the successor of the multiethnic Ottoman Empire, the much more ethnically homogenous Turkish Republic was still home to three large minorities of Kurdish, Armenian, and Greek origin in addition to other smaller minority groups, hence his point is not purely heuristic; Ahmad 1993: 682.

[62] Jacob M. Landau describes TIP as the "most durable" of Turkey's legal socialist parties, see Landau 1974: 122.

[63] Born in 1910, Aybar graduated with a law degree from Istanbul University and continued studying law in Paris. After returning to Turkey he practiced law and was outspoken against the Korean War.

[64] "Türkiye İşçi Partisi Büyük Kongresi" article published in *Eylem*, from the Kemal Sülker Collection, Box 604, of the International Institute of Social History in Amsterdam.

[65] Karpat 1967.

[66] Mango 2005.

Therefore, the program of the Türkiye İşçi Partisi was in many ways a continuation or broadening of the modernization project envisioned by Mustafa Kemal Ataturk rather than a clear break with it.

Beginning around 1963/64, the program of TIP took on an explicitly anti-imperialist character.[67] The general elections of 1965 provided an opportunity for TIP to galvanize opposition to the U.S. presence in the aftermath of the Johnson letter. The Workers' Party of Turkey publicized what was not intended to be public knowledge: the fact that the U.S. forces occupied 35.9 million square meters of Turkish territory. TIP's manifesto declared that "43 years after winning the War of Independence," it was necessary for Turkey to "regain its independence from the United States," which had established a presence in Turkey "under the guise of aid." Neither the Turkish police, nor Turkish judges, nor even Turkish military officers could enter the bases without permission. The bases represented "little Americas where the American flag is hoisted on Turkish land."[68] Although TIP had only been officially registered as a political party in 1961 and had held its first Party Congress in 1964, in the elections in 1965 it won 3 percent of the total votes and sent fifteen delegates to the parliament. These election results may not seem immediately impressive, but according to Landau these mere numbers may be deceptive: "In relation to the total Turkish population of over thirty millions in the 1960s, the number who actively participated in radical politics was small. However, like the drop of dye that suffuses the wool, it was they who colored the political life of the decade."[69] In a relatively short span of time, the newly founded party had acquired a social basis in the country and had created enough pressure on Prime Minister Süleyman Demirel to force him to respond to the issues that had been raised in the election campaign.

In December 1965, one of the first parliamentary activities of TIP was to submit a written inquiry to the government requesting a full list of all the documents signed between the U.S. and Turkey. This was tantamount to demanding an end to the secrecy surrounding the U.S. military presence and a first step in protesting against the collective disenfranchisement of the Turkish people by their American protectors. The first reaction of the government was denial. Demirel's response is still remembered today: "Üs yok, tesis var," in effect saying that there were no American bases, but merely "facilities."[70] Considering the size of the U.S. presence, it was unlikely that this verbal camouflage would be successful. Without any prompting, my interviewees have referred to this statement as "non-sense" and "comical."[71] A few months later, in April 1966, the Turkish government officially requested that the bilateral agreements be

---

[67] Interview with Akin Atauz in Ankara on July 30, 2006.
[68] Cited in Güvenç 2005: 91–92.
[69] Landau 1974: ix.
[70] "Demirel: Us yok, tesis var" in *Cumhuriyet*, as cited in *Cumhuriyet Ansiklopedisi 1961–1980*, Vol. 3, Istanbul, 3rd edition 2002.
[71] Interview with Oktay Etiman on June 26, 2006 in Ankara.

renegotiated. At first, Washington rejected the demand, but soon afterward agreed, although progress was very slow. It would take another nine months before the negotiations officially began in January 1967, and they would then drag on for more than two years. Soon Demirel had to publicly concede that he was attempting to revise the entangled gallimaufry of agreements.[72]

Merely reforming the conditions under which the U.S. military operated in Turkey, however, did not satisfy the demands put forth by TIP. During the Second Party Congress held November 20–26 in Malatya, the central issue was Aybar's concept of the indivisibility of the national-democratic and socialist revolutions.[73] In his address, Aybar argued that it was necessary to begin a full-fledged struggle against American imperialism as part of the struggle for socialism. His goal was not the implementation of reforms, but expulsion: "We are fully determined to engage in this struggle until such time as the last American soldier has left our country.... At that time victory will mean simultaneously the victory of socialism also."[74]

During the second half of the 1960s, forms of protest directed toward the U.S. presence gained momentum. As early as the spring of 1966, Aybar began describing the campaign against the U.S. presence as a second national liberation struggle, after the 1919–1923 War of Independence.[75] The comparison between the U.S. presence in Turkey at the time and the attempts by European powers after World War I to carve out spheres of influence in Anatolia was apparently not just a rhetorical slight-of-hand, but drew from a collective memory that was reactivated by the following events.

In November of 1966, a three-day riot erupted in Adana after a group of eight U.S. servicemen had allegedly harassed several Turkish women who were leaving a cinema. More than 2,000 people were involved in the rioting which resulted in considerable damage to U.S. military and civilian facilities in Adana. In what the Los Angeles Times referred to as a "mob outbreak," rioters sacked the Red Cross aid center, stoned the U.S. Consulate, damaged about fifty American-owned cars, and also broke the windows of the Turkish police station where the eight servicemen had fled after the violence began.[76] As this incident demonstrates, it was not just the policies of the U.S. government or American employers which caused resentment among the Turkish population, but also the behavior of officers and enlisted soldiers stationed there. Because the U.S. personnel enjoyed what essentially amounted to a form of diplomatic immunity, such outbursts of public rage were perhaps more effective in curbing misconduct on the part of G.I.s than the Turkish judicial system.

---

[72] *The New York Times*, November 12, 1965.
[73] "Türkiye İşçi Partisi ikinci büyük Kongressi" and "Basin Bülteni," November 20, 1966, from the Kemal Sülker Collection of the International Institute of Social History in Amsterdam.
[74] Lipovsky 1992: 22.
[75] Lipovsky 1992: 21.
[76] *Los Angeles Times*, November 15, 1966.

In his speech before the parliament in the aftermath of this incident, Life Senator Mehmet Özgüneş gave a speech in which he reminded the Americans that a "spontaneous resistance movement" started in southeast Turkey in 1919 against French troops after some of their soldiers were disrespectful toward Turkish women.[77] According to Oktay Etiman, who was born in Adana in 1947 and grew up there, the Americans who had been living side-by-side with Turkish families in the city of Adana for many years, henceforth began living in the military housing on the İncirlik base. In his own words: "the living together ended."[78]

Shortly after this event, in his New Year's address in 1967, Aybar made it clear that the Labor Party's campaign would be aimed at expelling the U.S. bases along with other forms of American influence: "I wish the Turkish people will have an independent Turkey rid of military bases, American soldiers, American experts and Peace Corps."[79] This focus on the U.S. military presence soon had an influence on the party's concept of socialism. According to Aybar, the most fundamental contradiction within Turkey was not that between worker and capitalist, but between U.S. imperialism with its "local landowner-comprador-pro US bureaucrat" and everyone else.

In addition to Aybar, Behice Boran was another outspoken TIP delegate. Boran criticized that although the military bases were considered "common" or "joint" defense installations, they did not represent the common interests of Turkey and the United States. She pointed out that the U.S. bases in Turkey were under the command of the Pentagon, and that even the Turkish defense minister had only restricted access to them.[80] By criticizing the fact that the bases were not administered jointly or on equal terms, she was attacking the very premise of NATO as an alliance. Boran further contended that the Soviet Union, having lost more than 20 million of its citizens in World War II, was far too weak to ever consider launching an invasion against Turkey. In the assessment of the TIP delegates, the Turkish elite had decided to join NATO to preserve its class position, not because of any real Soviet threat. In other words, NATO was more about preserving the capitalist system than about security. By questioning both the viability of the Soviet threat and the credibility of the U.S. presence, the İşçi Partisi had attacked the very essence of the American protection regime, namely its claim to offer legitimate protection against a truly dangerous outside threat.

Although their analysis was trenchant, and their rhetoric militant, TIP's strategy was less so. Aybar in particular supported "passive resistance" (*pasif direnme*) which meant refusing to engage in any form of contact with American

---

[77] *Turkish Daily News*, November 23, 1966.

[78] Interview with Oktay Etiman on June 26, 2006 in Ankara.

[79] Aybar speech reprinted in Ankara Daily News, January 3, 1967.

[80] National Assembly minutes, January 5, 1967, 29th meeting, 2nd session. Cited in Güvenç 2005: 93–94.

citizens in Turkey.[81] "Aybar called on traders not to supply Americans, on hairdressers not to style their wives' hair, and on ministry officials not to conduct talk with American specialists. In simple terms the 'national-liberation war' was to be waged peaceably and in accordance with the 1961 constitution." [82] Once the Americans realized that they were no longer welcome, it was assumed that they would not withdraw willingly, but would resort to force in order to stay, at which point the passive resistance would become active.

The dedication of the party leadership to passive resistance and parliamentary democracy, however, was not supported by all members of the party or other sympathizers. In an interview, Ertuğrul Kürkçü, the former chairman of the Revolutionary Youth Federation (Dev-Genç), described the effectiveness of this: "this was a pacifist way of protest, we don't have this tradition in Turkey: pacifism doesn't work in Turkey."[83] Within the Labor Party, a faction developed around Mihri Belli, who began propagating the concept of Milli demokratik devrim, or national democratic revolution (NDR). Belli's supporters were determined to take their cause to the streets rather than rely on incremental reforms through parliamentary means. According to Mihri Belli: "Aybar said, if we take our protests to the streets, fascism will come; we said, if we cannot take our protest to the streets, fascism *is* [already] there."[84]

The party became increasingly split for several other reasons. First, according to Lipovsky: "the most active and effective struggle against the American presence in Turkey was being waged by forces bearing the slogans of NDR,"[85] which increased the popularity of Belli's followers and led to further factionalism. Second, TIP did not organize among the student youth and hence did nothing to prevent left-leaning organizations from turning to armed struggle.[86] Third, the 1968 invasion of Prague by the Soviet Union caused fractions to develop within the Turkish left, as in many other countries. Therefore the split within TIP was further deepened.

It is important to emphasize that opposition to the U.S. military presence in Turkey arose before the escalation of the Vietnam War and before the formation of large anti-war movements in the U.S. or Western Europe. What triggered this opposition was not the war in Indochina, but rather domestic events in Turkey relating to the U.S. presence.[87] The escalation of the Vietnam

---

[81] "Genelge: Bütün İl ve İlçe Yönetim Kurulu Başkanliklarina," letter from Cemal Hakki Selek on October 14, 1966, from the Kemal Sülker Collection of the International Institute of Social History in Amsterdam.

[82] Lipovsky 1992: 46.

[83] Interview with Ertuğurul Kürkçü on July 4, 2006 in Istanbul.

[84] Interview with Mihri Belli in Istanbul on June 21, 2006.

[85] Lipovsky 1992: 73.

[86] Ibid: 79.

[87] For this type of argument, see "Turkey: A Time of Troubles" by Dwight J. Simpson in *Current History*, January 1972.

War and the Cyprus issue then contributed to increasing anti-U.S. sentiment in Turkey, but did not cause it.[88]

The 1969 elections were conducted on the basis of the majority-proportional system and not proportional representation; therefore, the votes for small parties were not counted. This new law was an initiative of the Justice Party (JP) to undermine small parties. Therefore, the failure of TIP was not because of abandonment by the voters but because of an attempt by the Justice Party to minimize the opposition.[89] By the end of the decade, the youth organizations had slipped out of control of the party and the rightist forces against TIP were gaining strength. On July 20, approximately four months after the March 12, 1971 military intervention, TIP was outlawed and most of the leaders of the party were arrested, including Behice Boran, Saban Erik, Said Ciltas, Sadun Aren, Osman Sakalsiz, Adil Ozkol and others.[90] Boran was sentenced to fifteen years in prison and Sadun Aren to twelve years.

A July 11, 1970 report ("Summary of Current Status of Leftist Movements in Turkey") was sent from the embassy to the Department of State, outlining how the legal left had been weakened while the revolutionary left had been strengthened. About the fate of the Labor Party it wrote: "The TLP is at present in its weakest state since its inception in 1963, and the Turkish political system may eventually suffer from the weakening of this important safety valve."[91] However, the decade did not end in total defeat. Ironically, although TIP was divided, outlawed, and its leaders imprisoned, some of their most central ideas lived on within both larger, more established parties such as the Republican People's Party (CHP) and within the student movements, which soon took on a life of their own.

Opposition in the Street: Protests against the Sixth Fleet

There were numerous organic connections between TIP and student activists in the first half of the 1960s. The first "idea club" (*fikir kulub*) was founded in 1956 by a group at the University of Istanbul. Beginning in the early 1960s, the number of idea clubs increased, which led to the creation of an umbrella organization, the Federation of Idea Clubs (FIC). In 1967, the FIC accepted

---

[88] This does not mean, of course, that the Vietnam War was insignificant. On the contrary, Aybar himself was a member of the International War Crimes Tribunal founded by Bertrand Russell in 1966 and held over two sessions in 1967 in Stockholm and Copenhagen. Other members of the tribunal included Jean-Paul Sartre, Simone de Beauvoir, James Baldwin, Stokely Carmichael, Peter Weiss, and Isaac Deutscher.

[89] Hale 1994: 177.

[90] See Harris 2002, Lipovsky 1992, or Landau 1974 for a discussion of the impact of the 1971 military intervention on the Turkish Labor Party.

[91] "Current Status of Leftist Movements in Turkey" from Embassy to Department of State, July 11, 1970, Record Group: 59, Stack Area 150, Row 67, Compartment 20, Shelf 1-2, Description: Subject Numeric Files 1970–1973, Box 2638.

the strategy of the Labor Party concerning the indivisibility of socialist and national-democratic goals in Turkey.[92] Although the student federation agreed with the theoretical program of TIP, it did not agree with its tactics; this became especially clear during the visits of the Sixth Fleet to Turkish ports. Before discussing these events, however, it is necessary to make a few remarks about naval power and the Mediterranean.

During the Cold War, the Mediterranean Sea was one of the most heavily militarized bodies of water in the world and in many ways the ebbs and flows of détente and confrontation between the superpowers could be read according to the types of ships that were plying its waters. As the Cold War developed, so too did the military presence in the Mediterranean. Two years after the visit of the USS Missouri to Istanbul, the Sixth Fleet was formed in June 1948 with the official purpose of supporting the Truman Doctrine. Although the fleet was later assigned a NATO mission, it has essentially always been a unilateral effort in naval diplomacy and power projection on the part of the United States.

One of the major turning points in U.S. Naval history during the Cold War was when the Sixth Fleet was used to pressure the British and French to stop their attempt to reoccupy the Suez Canal in 1956. In the aftermath of this confrontation between the United States and the two largest remaining colonial powers, the British abandoned their huge base at the Suez Canal. This is often regarded as marking the beginning of the end of Britain's ability to take independent action in the realm of foreign policy. At this point, the Sixth Fleet become solely responsible for the defense of the Southern Flank and the absence of any other presence increased the Fleet's prestige as the sole guardian of the Mediterranean.

After the Cyprus crisis of 1964, however, Turkish-Soviet relations improved. This was followed by a deployment of Soviet naval forces through the straits to the Mediterranean, marking a challenge to absolute U.S. hegemony in what had hitherto been called an "American Lake." By the fall of 1968, the Spanish and French were discussing the possibility of a "neutralized Mediterranean."[93]

In the late 1960s, the Sixth Fleet consisted of more than 4,000 sailors stationed on two attack carriers with more than 150 combat aircraft, escort ships, nuclear submarines, an amphibious task force, and supportive service ships; therefore, when it docked for rest and relaxation, it was a major event for most cities. Although these port calls had been welcomed for more than twenty years as a source of revenue for local businesses, this began to change in the late 1960s.

Port visits varied in length, but usually lasted several days or up to a week. According to a Naval Dispatch Summary from June 27, 1967, the "Anti-US/

---

[92] Lipovsky 1992: 118.
[93] "The U.S. Sixth Fleet: Search for Consensus" by Lieutenant Commander P. A. Dur in U.S. Naval Proceedings, June 1974, Vol. 100, No. 4/856.

Sixth Fleet protest march" attracted around 500 "Turkish student demonstrators" as well as 1,500 "spectators and sympathizers." Because the demonstration took place on the fleet landing, sailors were prevented from leaving the ship until the protest was over.[94] A secret memo from the U.S. Consulate in Istanbul to the Secretary of State in Washington, DC regarding the visit on June 24, 1967, describes the demonstration and its implications:[95]

Rather than prevent forcibly any demonstrators from molesting fleet personnel, Istanbul military commander Maj. Gen. Selami Pekun requested all fleet personnel withdraw to ships. Implications: visits approved by TGS are to be interrupted if leftist students do not approve; visitors safe on shore only when not threatened and if threatened, must withdraw. Turkish Army officers allowed selves be carried on shoulders of demonstrators against allied fleet. No need point out implication.[96]

During a subsequent visit of the Sixth Fleet in October 1967, activists distributed a flyer that compared Istanbul to Saigon, where the Seventh Fleet was docked, to draw parallels between the U.S. intervention in Vietnam and the U.S. presence in Turkey. The flyer argued that Istanbul, once the capital of the Ottoman Empire, had been reduced to a "Center for Sexual Relaxation" for the U.S. Navy and included pictures of U.S. sailors attending a nightclub and belly dancing show.[97]

During a visit of the Sixth Fleet the following spring in April 1968, some of the demonstrators went a step further and rather than merely passing out flyers, tried to interact with the sailors by stealing their caps when they disembarked from their ship and then tossing the caps to and fro while the sailors tried to retrieve them. According to two participants in the protest who were interviewed separately, the atmosphere of the event was "playful" as there was no attempt to harm any of the U.S. sailors or even intimidate them.[98] The U.S. civilian and military authorities in Turkey, however, apparently perceived things differently; they became alarmed by the events and there were numerous memos sent from the American Consulate in Istanbul to the American Embassy in Ankara alerting them of the situation. In these memos, the American side accused the Turkish police of not adequately providing for the safety of the U.S. sailors:

Conclude that GoT [Government of Turkey] may have decided it cannot permit pictures of police arresting Leftist students protesting presence [of] fleet because of possible

[94] Sixth Fleet Weekly Dispatch Summaries from the U.S. Naval Archives.
[95] I have copied the memos exactly as I found them in the National Archives and Records Administration in Washington, DC; therefore, any mistakes or short-hand is in the original document.
[96] The following documents can be found in the National Archives and Records Administration (NARA) in Washington, DC, unless otherwise indicated.
[97] Memo from the American Consulate in Istanbul to the Embassy in Ankara, October 7, 1967.
[98] Interview with Masis Kürkcügil in June 2006 in Istanbul; interview with Akin Atauz in Ankara on June 30, 2006.

leftist charge GoT behaving like US Govt which leftist press has been vilifying since King assassination.[99]

In a further memo from the Embassy in Ankara to the Secretary of State in Washington, DC on May 1, 1968, the predicament in which the Turkish government found itself was expressed with a clarity that is only reserved for internal debates:

As member of NATO Turkey has duty to make provision for personnel of sixth fleet to visit Turkish ports under peaceful conditions. However, Turkey is a democratic country and cannot prevent demonstrations.[100]

The Turkish government was therefore caught in a contradiction between its obligations to the United States as a NATO ally and its obligations to its own constituency as a democratic state. In this sense, it was under pressure from both above and below and had to cooperate with demands from Washington while not appearing to be a client of the United States.

Despite these tensions, the Navy continued to make "friendly calls" to major Turkish port cities. Between July 17 and 23, 1968, the Sixth Fleet, including a number of ships such as the carrier Independence and several destroyers, anchored in Istanbul. On the first day of shore leave, the soldiers were greeted by a small group of students carrying signs bearing the inscription "Beat it, Fleet" and "NATO: No." By the end of the week, however, a similar anti-U.S. rally attracted more than 5,000 people and demonstrations or acts of vandalism directed at U.S. institutions had spread to Ankara, Izmir, Trabzon, and Konya. In what the *Turkish Daily News* referred to as a "wave of anti-Americanism," Turkish youths threw black paint and ink onto U.S. sailors, besmearing their traditional white uniforms while others swung clubs. A number of American institutions such as the Turkish-American Bank, the USIS center, the American Library, and the Pan-American Bank all had their windows broken.

At the end of the week, more than twenty American sailors had been wounded and one Turkish student had been killed by the Turkish police. The police, which had obviously changed its stance toward the demonstrators, not only violently disrupted the protest but chased some of the students to the boarding house where they lived. This led to further clashes between Turkish youth and the police when more than 2,000 students delivered an empty coffin to the door step of the Justice Department to protest police brutality.

The theme of sexual exploitation was also addressed by student leaders. The President of the Student Union at Istanbul Technical University, Harun Karadeniz, held a press conference and declared "Istanbul is not the brothel of

[99] From the U.S. Consulate in Istanbul to the Embassy in Ankara; April 10, 1968.
[100] State Department Memo from U.S. Embassy in Ankara to Secretary of State Washington DC, May 1, 1968.

the Sixth Fleet. We will continue our fight against imperialism."[101] Some student leaders accused the Turkish police of acting under order of the U.S. military authorities. Whether or not this is true, the fact that the Turkish police had killed a Turkish student was in many ways a defining moment for the left: it was the first time that such brutality had been used to crush the student movement. Additionally, these events did nothing to increase the prestige of the Sixth Fleet in the eyes of the Turkish population, as becomes clear in the following editorial in the *Turkish Daily News* from July 27: "The visits are meant to be a show of strength. But the sailors are under orders not to hit back and must rely on the protection of Turkish police. The sight of American sailors cowering under police protection or leaping into the water to escape a club-swinging mob does not reinforce ... U.S. strength."[102]

The U-2 spy plane incident, the missile crisis, and the Johnson letter had already made U.S. protection appear questionable. The killing of a student who was protesting against the Sixth Fleet was, in the eyes of the student movement, not only an unforgivable act, but illustrated the Janus-faced nature of a protection regime that harms those it claims to protect. As the U.S. presence increasingly caused collateral harm to the non-U.S. citizens it aimed to protect, the transformation from legitimate to pernicious protection had begun.

Yet even these dramatic events did not immediately sway the decision-makers in Washington. Already in a joint telegram from the State Department and the Department of Defense in April of 1968, it was suggested that the "NATO theme" should be stressed and that "an occasional public statement or welcome from high Turkish authorities would be helpful, as would greater Turkish military association with such visits."[103]

These suggestions were duly implemented by Ankara during the following Sixth Fleet visit in Izmir during late August and early September of 1968. President Sunay's radio speech in which he welcomed the Sixth Fleet was played a full three times on August 30, 1968, which, somewhat ironically, is the Victory Holiday in Turkey that celebrates the Turkish military's expulsion of the Greek forces. Perhaps more importantly, Turkish Navy ships sailed into Izmir with the U.S. ships and anchored alongside them. Turkish sailors also accompanied the U.S. sailors throughout their stay in Izmir.

Because the Sixth Fleet visits were relatively quiet compared to the events in Istanbul six weeks earlier, they were considered a success. The Turkish Daily News made sure that its readers understood this in a front-page article:

---

[101] *Turkish Daily News*, July 20, 1968. The TDN was the main English-language newspaper that was published in Turkey at the time. It appeared six days a week and covered events involving the U.S. personnel in Turkey; it is highly likely that many of its readers were either part of the diplomatic corps or U.S. military. However, because it was not an official military publication like *Stars and Stripes*, it can be assumed that it was more objective in its portrayal of certain events.

[102] *Turkish Daily News*, July 27, 1968.

[103] Memo from April 15, 1968.

[B]ig-scale Sixth Fleet visits can take place without incident. This has been proven. Now the visits, which officials realize are an irritant to sensitive Turkish nationalist feelings, can be cut back, cut out, or cloaked in a NATO mantle without communist claims that anti-Americanism in Turkey had forced the changes.[104]

An Admiral of the Turkish Navy, Sezi Orkunt, had a different interpretation of the Izmir visits. He explained the significance of the events in an article entitled "Visits of the Fleet" published in *Outlook* on September 4, 1968:

Because the naval forces bear the most recent and deadly innovations in modern weaponry, they are a mobile weapon fair bringing different thoughts to the minds of friends and foes and are a silent but impressive media of pressure.

Despite their inclination to cancel the port visits because of the recent tensions, Admiral Orkunt argues that the Turkish officials were forced to agree to the visit because the U.S. insisted that it was "normal" for allies to make port calls. Orkunt calls this a "needless provocation" and argues that the Turkish government should be able to decide itself how to deal with the opposition to the visits and whether or not they should take place at all.

America has no right seeing and pointing out to the Turkish youth as "extreme left" in the same parallel as "communist." ...We also know how to iron out the extreme ends. The Sixth Fleet's desire to and insistence in coming to Izmir is the result of an exclusively American point of view.[105]

Whether Admiral Orkunt was speaking for himself or on behalf of a larger contingency cannot be determined with certainty. His article, however, is an indication that opposition to the Sixth Fleet visits had spread from youth activists to the upper echelons of the Turkish Navy.

February 1969 saw the largest and most pluralist anti-Sixth Fleet protests yet, with more than 30,000 people taking part in the demonstrations. In addition to the students, trade unions, workers, and opposition parties had joined together in an alliance to demonstrate that the warships were not welcome. Most major newspapers opposed the visit and even otherwise nonpartisan bystanders had begun to question whether the port visits were worth the trouble they caused. Opposition to the navy was no longer the purview of small leftist student groups, but had become commonplace within the political mainstream. However, right-wing commandos had been organized to destroy the unity of the movement. Although the demonstration on Taksim Square was peaceful, neo-fascists and Islamists attacked the demonstrators with stones, iron bars, and butchers' knives. Three people were killed and more than 200 were injured. Radio Istanbul interrupted programs to appeal for blood donors.[106] February 16, 1969, became known as Bloody Sunday (*Kanlı Pazar*)

---

[104] *Turkish Daily News*, September 10, 1968.
[105] "Visits of the Fleet" by Admiral Sezi Orkunt in *Outlook* on September 4, 1968 in: Air Force Microfilm Reel.
[106] Cohen, Sam, "6th Fleet Stirs up Turk Resentments," *The Washington Post*, February 17, 1969.

and came to symbolize the growth of right-wing violence to counter the growing hegemony of the Turkish left.[107]

Opposition on the Bases: Strikes of Turkish Workers
at U.S. Military Facilities

The various types of protest against the U.S. presence discussed thus far resulted
in what can be considered "external operational disruption"; now I turn to
unrest on the U.S. bases themselves. The strikes of base workers who were
employed on the various military facilities in Turkey resulted in "internal operational disruption," a potentially much more explosive form of social unrest as
it threatened the very functioning of the protection regime.

In 1947, the CHP officially passed legislation (Law 5018) that allowed
union formation. At the same time, the activities that unions could participate
in were curtailed: unions were not allowed to strike or to engage in politics.
The democratic opening with the emergence of the multi-party system in 1950
did not immediately lead to greater democracy for union organizations.[108] In
fact, there was a certain amount of continuity in terms of the repression of
independent working-class organizations. The 1950s did witness, however, the
development of the first national confederation of unions, *Türk-İşçi Sendikası
Konfederasyonu* (Türk-İş). From the perspective of the U.S. government and
the American Federation of Labor (AFL), a national confederation would make
it easier to influence the development of the Turkish labor movement. Irving
Brown, who served as the AFL's director in Europe during the early postwar
period, was present at the meeting in April 1952 which led to the founding of
Türk-İş. By this time, both the major political parties in Turkey (CHP and DP)
seemed to have accepted the American perspective that labor militancy could
be held in check through the confederation.

After the constitutional changes that took place in the wake of the 1960
coup, the founding of the Workers' Party (TIP), and several major demonstrations, unions were finally granted the right to strike in 1963.[109] Instead of
throwing its weight behind the Workers' Party, which was the most outspoken
in its defense of working-class rights, Türk-İş passed a resolution in early 1964
that it would remain "above party politics." The decision to remain apolitical
was not welcomed in all quarters of the labor movement, and by 1967 a rival

---

[107] These events soon became part of the collective memory. In February 1971, a theater company
staged a production of Hamlet in a working-class neighborhood of Istanbul. Shakespeare's
drama, however, had been adapted to the current political situation in Turkey: rather than the
Prince of Denmark, Hamlet was a leftist student leader protesting the Sixth Fleet. Video footage
of the Bloody Sunday demonstration was played in the background. Memo from the Consulate
in Istanbul to the State Department on February 5, 1971.

[108] For a discussion of this period, see Mello 2008.

[109] Mello 2008: 219.

confederation had emerged: the Confederation of Revolutionary Labor Unions *(Devrimci İşçi Sendikası Konfederasyonu* – DİSK).

Turkish workers employed at U.S. military installations such as TUSLOG and JUSMAT were and are still organized in the Federation of Defense Industry and Allied Workers Union (Harb-İş), which is a member of the older, more mainstream Türk-İş confederation. The total membership of Harb-İş in the late 1960s was 42,000, with 34,000 of these workers employed by the Turkish military and approximately 8,000 working on the U.S. bases. According to Kenan Durukan, who was President of Harb-İş from 1971 until 1993, at that time 100 percent of all military employees were organized in the trade union.[110] Beginning on June 30 1967, Harb-İş became involved in a labor dispute against the Tumpane Company in Ankara and Adana. However, because the strike was scheduled to coincide with the joint NATO exercises in Merzifon, the dispute evolved into much more than a matter of higher wages and became symbolic of the tensions inherent in the U.S. presence in Turkey. Although the Turkish government tried to prevent an escalation by postponing the strike for thirty days on the grounds that it endangered national security, the strike eventually erupted on September 22 in Adana at the İncirlik Base and then on September 25 in Ankara. The sites run by the Tumpane Company that were affected by the strike include the American school and air station in Balgat, Maltepe repair shop, Tuslog, Merhaba Palas Hotel, the American billet, and JUSMATT. According to reports in the *Turkish Daily News*, there were a number of "nasty incidents" between the Turkish workers who were on strike and U.S. military personnel: "The American military authorities are accused of driving cars on the strikers, attacking them with iron bars, threatening them with specially trained dogs and soldiers carrying Thomsons."[111] According to other articles from the *Turkish Daily News*, strikers interfered with the movements of Americans on and off base, scattered garbage on the grounds and streets surrounding İncirlik, and turned off power to water pumps inside the İncirlik facility.[112] During the labor dispute, sixteen cars belonging to Americans in Ankara had their tires knifed. According to an announcement by Harb-Is, "Our struggle has been more with the American military authorities than with Tumpane." The labor dispute was resolved in October and resulted in a 20 percent wage increase for the employees of Tumpane.[113]

A second strike erupted less than two years later, in the spring of 1969. Although the second strike was not scheduled to disrupt NATO exercises, it did coincide with the final phase of bilateral negotiations regarding the status of U.S. forces in Turkey; therefore, it became equally if not more contentious. The labor dispute began with a walk-out at the U.S. facility in Izmir on April

---

[110] Interview with Kenan Durukan in Istanbul on June 20, 2007.
[111] *Turkish Daily News*, September 30, 1967.
[112] *Turkish Daily News*, October 2, 1967.
[113] *Turkish Daily News*, October 6, 1967.

17 and union demands for higher pay and a say in the hiring and firing of
Turkish workers. The strike soon spread to Istanbul, Iskenderun, Ankara, and
Adana. Because the base workers often benefited from salaries that were higher
than those that were paid for comparable jobs off-base (for example, a driver
for the Turkish Ministry of Defense earned $94 a month, whereas a driver
for the Americans with the same seniority earned $158), it is unlikely that the
primary grievance of the employees was their pay rate.[114] In fact, the Turkish
Employers' Association often put pressure on Harb-İş because the high wages
paid at the U.S. facilities were causing workers in other sectors to demand
higher wages as well.[115] Instead, it would seem as though workers' demands to
have more say in the hiring and firing of Turkish employees was in line with
demands by actors in other social movements to increase Turkish control over
the U.S. presence.

More than three weeks into the labor dispute, Frank E. Ruggles, a major in
the U.S. Army stationed in Izmir, wrote a letter to his representative in the U.S.
Congress, demanding that some action be taken. Below is an excerpt of the
letter dated May 13, 1969:

Three weeks ago a local Turkish labor union struck against the Tumpane Company,
the US support company in this area, forcing the closing of the exchange facility, the
commissary, news stand, movie theater, transient military hotel and officer and enlisted
messes; similarly affected were facilities at a nearby US Air Force activity where only
the theater remains open. A week later the union struck against US Military transportation and liaison activities. The first strike act cut us off from approved food services
and other necessities of health and welfare. The second act has totally restricted the
movement out of Turkey of automobiles and household goods.... I will be leaving this
country shortly for assignment to Vietnam and because of this situation I find myself in
a most uncomfortable and precarious position, for surely I cannot leave here without
my personal belongings.... The support we can normally expect as military personnel
serving overseas has vanished.... The Turkish police have stood by and done little in
the face of rioters and vandals, nor have they prevented the rioters from mounting their
attacks.[116]

Ten days later, the strike situation seems to have further deteriorated. In a
joint message from the Department of State and Defense Department to the
American Embassy in Ankara on May 23, the concerns expressed by Major
Ruggles were underlined:

We are therefore increasingly concerned with health and welfare of US personnel as
well as threat to their safety as now posed by illegal actions and activities of strikers.
It appears to us that efforts of Turkish authorities to see peaceful and orderly strike

[114] "Turkish Strike, by Blocking US Supplies, May Speed American Cutback," *The New York Times*, May 7, 1969.
[115] Interview with Kenan Durukan in Istanbul on June 20, 2007.
[116] NARA archives.

conducted in strict compliance with Turkish labor laws continue to be inadequate and half hearted.[117]

The same day, the *Chicago Tribune* reported that the "exodus" of dependents of U.S. military personnel had reached the 2,400 mark. Although there was no official decision to remove dependents from Turkey, many families apparently took the initiative themselves and had their wives and children flown out of the country, often to neighboring Greece. The same article raised the specter that the United States had lost control of its facilities with its opening sentence: "About 1,400 striking workers have assumed effective control of all five American-run NATO bases in Turkey."[118]

To what extent the United States feared that it had indeed lost control of its bases in Turkey is unclear, however, it is clear that the strike was becoming a serious matter in the United States, as even the U.S. Congress began to get involved in the labor dispute. In particular, the strike appeared to be threatening the resolution of the new bilateral agreement between the two NATO allies. On May 29, Senator J. W. Fulbright representing the Senate Armed Services Committee wrote a letter to the Secretary of State, William P. Rogers, sharply criticizing the fact that the Armed Services Committee was not involved or even informed of the negotiations, although they had been going on for two and a half years. They were finally informed on May 22 (at which point the strike was in its fifth week) and the administration hoped to have the new agreement signed on May 24, two days after the hearing in the Senate. Senator Fulbright then addresses the urgency of the situation:

[U]nless the Turkish Government – which apparently wants this agreement – is prepared to accept language that gives us some *protection against strikes* at these facilities, we will have lost any chance for a reasonable settlement of what looms as our immediate chief problem in that country. (emphasis added)

On May 30, after six weeks, the strike was finally settled with the union winning most of its demands. The 16 percent wage increase was retroactive to October 1968 and the fringe and social benefits were retroactive to January 1. Most importantly, perhaps, the Turkish people "feel they have won a moral victory over the Americans" which was considered especially significant in the run-up to the October 12 parliamentary elections.[119]

In June, a country team undertook a study of the effects on U.S.-Turkish relations caused by the strike. In their attempt to draft a well-balanced report, in addition to the negative consequences of the strike such as the impact on Congressional and public opinion in the United States, and an increase in anti-

---

[117] NARA Archives; State Department – Central Foreign Policy Files.
[118] "Report Turks Take Control of NATO Bases: Anti-US Sentiment Seen as High," *Chicago Tribune*, May 24, 1969: "
[119] "Strike Leaves US-Turk Ties Snarled," *Christian Science Monitor*, June 4 1969.

Turkish and anti-U.S. sentiment resulting in strained relations on both sides, one positive outcome was also mentioned:

On plus side, [Government of Turkey] regarded strike as our problem at outset, but gradually came to realize importance to itself of secure American operations.... We believe [Government of Turkey] now seized of need prevent repetition, partly because they see violence against US as part of general problem of disturbances in Turkey today and partly from apprehension over repercussions in US. We expect GoT cooperation in forward contingency planning.[120]

The U.S. media was somewhat less even-handed in its portrayal of post-strike bilateral relations. In an article titled "Strike Leaves US-Turk Ties Snarled" the *Christian Science Monitor* summarized the violence: "Turks stabbed Americans. Americans punched Turks. US Air Force Exchange trucks carrying food were hijacked. Turkish teen-agers reportedly shook lye powder in the face of the 10-year-old child of an American sergeant, and school buses were sabotaged." Whereas the country team report indicated that the strike led the Turkish government to recognize their mutual interests, media reports emphasized that the Turkish government had difficulty emphasizing any mutual interests in public. "The strike had made it an even greater political risk than before for Premier Demirel, or any other future Turkish government, to publicly stress cooperation with the US." It would seem that Turkish officials wished that whatever cooperation did exist, that it would be less obvious: "Turkish commentators acknowledge the effort made, but not the success of the effort, to give the American presence here what Washington calls a 'low profile' – that is, unobtrusiveness."[121]

Finally, the strike led the U.S. authorities to demand changes in Turkish labor legislation. After having supported the introduction of modern labor-management relations a few years earlier, the United States was now more interested in restricting the rights of organized labor. The first priority of the American side was to change the legislation so that U.S. military facilities would receive the same "protection" from labor strikes as Turkish military facilities.

Paragraph II, Article 20 of Turkish Labor Law 275 provides that it is unlawful to call a strike or order a lockout in an establishment run directly by the Ministry of National Defense, except where the Central Court of Arbitration finds that the wages or other conditions are less favorable than those in similar establishments in other sectors. If this same law were applied to U.S. installations, strikes would become illegal.[122] However, in a letter from the Secretary of State in Washington, DC to the embassy in Ankara, the government of Turkey was accused of "legalistically and bureaucratically" arguing that "this is

120 Report from the American Embassy in Ankara to the Secretary of State in Washington, DC, sent on June 18, 1969.
121 "Strike Leaves US-Turk Ties Snarled," *Christian Science Monitor*, June 4 1969.
122 TUSLOG report of Strike, to Headquarters U.S. Air Force, Air Force Archives, microfilm reel 3777.

impossible because these provisions, in words of law, apply to sites run directly by the MOD (Ministry of Defense)."[123] The letter continues:

Turks have responded that they could not (repeat) not manage this prior to elections and have urged US to proceed with signature basic agreement on basis oral assurance they would do what they could after elections. Our painful, costly, damaging – and indeed dangerous – experience in repeated labor disputes in Turkey has made it crystal clear that *we must have the protection we seek*. Otherwise both USG and GoT vulnerable to having their national security endangered by strikers, as in 1967 when they came close to cutting off power to ... İncirlik, (have just been informed İncirlik now closed due lack of generator fuel).... What is involved is not (repeat) not what Kuneralp refers to "irresponsible acts of a few workers," but irresponsibility on part Turkish authorities charged with enforcing the law. They, perhaps understandably, have no stomach for cracking down on their own people in defense of foreigners – especially in an election year. (emphasis added)

As in the situation during the Sixth Fleet protests, the U.S. officials did not just see the problem as being the result of the students who were protesting or the workers who were striking, but instead saw the more significant problem as being that the Turkish authorities were not sufficiently protecting U.S. personnel. This, apparently, is what led to suspicions that the elements within the Turkish state were no longer on their side.

Another suggestion that was discussed was to Americanize by hiring U.S. citizens, especially for key positions such as the middle-management positions and drivers who delivered important supplies to the bases. The problem with this solution, however, was that this would be considerably more expensive and also contradicted the intention to keep the number of personnel to an absolute minimum. The lack of Turkish language skills on the part of many Americans also posed a problem in this regard. For example, at a bare minimum it was necessary for drivers to be able to read the road signs in Turkish. A third suggestion was to have workers hired directly by the Turkish Ministry of Defense and only indirectly through the United States and its subsidiary, the Tumpane company. That would make the Turkish government responsible for labor issues and would, it was hoped, force them to take a more active role in maintaining order. This, however, also posed problems because the Americans knew that the Turkish government would, in the event of a strike, distinguish between mutual defense facilities and support facilities. According to a letter sent to the Chief of Staff of the U.S. Air Force from the Headquarters of the U.S. Air Force in Europe in June 1969, the government of Turkey sees the snack bars, bowling alleys, retail stores, officers clubs, and so on as the "least desirable visible evidence of the US presence in Turkey" and hence these facilities would most likely not be protected during a strike.[124]

[123] Letter from Secretary of State to the Embassy in Ankara, May 1969, Air Force Archives, microfilm reel 3777.
[124] Letter from the Chief of Staff of the U.S. Air Force to the Headquarters of the U.S. Air Force in Europe in June 1969, Air Force Archives, microfilm reel 3777.

Obviously, there was no easy solution to the problems caused by the strikes. The Labor Party laid the groundwork for questioning the U.S. presence and created an opening in Turkish politics. By demanding that the secret agreements be made public, TIP was essentially demanding an end to the disenfranchisement of the people by the U.S. presence. The protests against the Sixth Fleet visits went a step further in that they disrupted the actual operations of the navy. Activists had the power to do this precisely because the U.S. Navy relied on the infrastructure of its host country, in this case the ports and harbors of Istanbul and Izmir. I have termed this a form of external operational disruption. Civilian base workers, however, also had the power to disrupt the internal operations of the U.S. military presence – because the United States was dependent on their labor power. The strikes, therefore, were perhaps the most disturbing type of protest as they resulted in internal operational disruption and severely disrupted the ability of the United States to operate in Turkey. I return to the question of the changing U.S. profile in the concluding section. Ironically, the U.S. military – which claimed to be protecting Turkey from the Soviet Union – found it necessary to demand "protection from strikes." As demonstrated in the following section, they soon faced the threat of violent attacks for which they also required the protection of the Turkish police.

Opposition to the U.S. Presence and the Turn to Armed Struggle

The year 1969 was decisive in several ways. In addition to the strike, the new election regulations had made it very difficult for small parties to enter parliament, leading some on the left to believe that change had to be pursued in other ways. Additionally, the collective experience of Bloody Sunday was interpreted in different ways: whereas some were undoubtedly intimidated by the appearance of right-wing commandos on the scene, others came to the conclusion that violence could only be answered with violence. Although the majority of the demonstrators in the Sixth Fleet protests did not intend to harm any of the individual sailors, this cautious, hands-off approach was not applied to the protests against the arrival of the new American ambassador, who was seen as more complicit in U.S. foreign policy than the enlisted sailors. One event in particular exemplifies the changing protest culture.

On January 6, 1969, Robert Komer, the new U.S. Ambassador to Turkey, paid a visit to the Middle East Technical University (METU) in Ankara. Prior to his appointment he had served in Vietnam as the "Chief of Pacification." The so-called pacification program was intended to "win the hearts and minds" of the South Vietnamese. In other words, it was essentially the propaganda component of Operation Phoenix, which, according to official testimony, had resulted in the death of more than 20,000 Vietnamese. The announcement that Komer would be the new U.S. Ambassador to Turkey, a man whose reputation had earned him the nickname "Blowtorch Bob," immediately set off demonstrations.[125]

At that time, METU was in some ways an autonomous zone for the left. As the 1960 constitution barred the police from entering university premises, many of the student groups had set up their headquarters on campus, where they could operate relatively freely. Hence, when students discovered that the U.S. ambassador had entered the campus and was paying a visit to the rector of the university, this was taken as an act of trespassing and outright provocation. A group of approximately 300 to 500 students gathered and began shouting outside the office of the rector of the university, Kemal Kurdaş, who was meeting with Komer. When their shouting failed to result in any response from the rector's office, the crowd of students decided to attack the automobile with which Komer had arrived on campus. A group of ten to twelve students began shaking the car while the others cheered them on; soon the car was overturned and set ablaze.[126] What is perhaps even more remarkable than this spontaneous act on the part of the students, is the reaction by Kemal Kurdaş, the rector, who did not call the police even after the car was in flames. According to Akın Atauz, who was involved in the event and spent several months in prison as a result: "This would have been like calling the enemy into your homeland" – such was the pressure on him created through the student protest culture.[127] The message sent by this event did not fall on deaf ears. In a report prepared for the Committee on Foreign Relations in the U.S. Senate in 1980 – eleven years after the Komer incident – the burning of his car is mentioned as an example of student radicalism.[128] However, rather than identifying U.S. foreign policy as one of the causes of such student unrest, the rapid increase in student enrollment is blamed (METU went from an enrollment of 40 in 1956 to 13,000 in 1979), even though this was one of the goals of the modernization of the educational system in Turkey that the United States explicitly supported when it contributed to the founding of METU through financial grants. The report identifies two sources of "extremism" – one of them being the overpopulated *gecekondu* housing tracts, and the other being overpopulated university campuses.

The dismissal of Ambassador Komer in May 1969 after only four months in office because the uproar involving his appointment, perhaps also reaffirmed

---

[125] Approximately 750 students showed up at the airport on November 28, 1968 to protest Komer's arrival, calling him the "Executioner of Vietnam" and the "Butcher of Vietnam." Apparently to avoid the protesters, Komer's plane debarked at the far end of the runway. Source: History of TUSLOG Det. 170, July 1, 1968 – December 21, 1968, Air Force archives, Roll number M 3037.

[126] The fact that the African-American chauffeur was allowed to leave the car unharmed was explained to me as evidence that the students distinguished between Americans of different racial and class backgrounds. They wanted to direct their rage specifically at Komer as representing the foreign policy establishment of the United States. Interview with Akin Atauz on June 30, 2006 in Ankara.

[127] Interview with Akin Atauz on June 30, 2006 in Ankara.

[128] Binnendijk 1980.

the belief that confrontational tactics would produce results. In the fall of 1969 the Revolutionary Youth Organization (Dev-Genç) established itself as a type of umbrella organization for other leftist groups who shared their belief that armed force was necessary to take control of the state apparatus. Two groups which stemmed from Dev-Genç were the Turkish People's Liberation Army (THKO), led by Deniz Gezmiş, and the Turkish People's Liberation Front (THKC). Because the THKC was closely associated with another group called the Turkish People's Liberation Party (THKP), the two groups were soon indistinguishable and referred to jointly as THKP-C.[129] Mahir Çayan, leader of the THKP-C during the early 1970s, urged members of the group to go beyond campus activities and work to unite Kurds and Turks in a common struggle. The THKO under Gezmiş was more oriented toward guerrilla activities in the mountains and rural areas.[130]

Both CIA reports and interviews with former members of these groups have confirmed that there were significant connections between these Turkish groups and Palestinian organizations such as Al-Fatah. The years 1969 and 1970 saw the beginning of a wave of violent attacks against American personnel. After a flurry of warnings sent by the U.S. embassy and consulate back to the State Department warning them of the rising tide of violence, it was decided that Under Secretary Elliot Richardson should discuss the matter with Mr. Melih Esenbel, the Turkish Ambassador in Washington. Mr. Esenbel was urged to ask his government to do more to protect U.S. personnel and property, because otherwise if the situation continued it could "put a severe strain on US-Turkish relations."[131] At the same time as pressure was being put on Ankara to clamp down on the terrorist activity, other U.S. officials were urging high-ranking Turkish military commanders to show restraint and not intervene in the political cauldron. Frank Cash, the Country Director for Turkish Affairs discussed the escalating violence with General Muhsin Batur, Commander of the Turkish Air Force in mid-January 1971. Mr. Cash argued that it would be a "great tragedy" and "grave setback" if the Turkish military would take over, because Turkey would lose its "rare status" as an economically developing democracy, which could endanger the flow of U.S. military and economic aid.[132] While the messages emanating out of Washington continued with a certain degree of inherent contradiction, the perpetrators of the violence became more daring in their tactics.

In November 1970, two U.S. installations in Ankara were attacked by bombs. In the predawn darkness of December 29, two Turkish policemen guarding the American embassy were killed, receiving multiple bullet wounds.

---

[129] For a more detailed discussion of the THKO, THKP-C, as well as other groups engaged in armed struggle in Turkey, see Ulus 2011.

[130] Interview with Oktay Etiman on June 26, 2006 in Ankara.

[131] Memo prepared for the Under Secretary by the NEA from Joseph J. Sisco on April 29, 1970 for a meeting that he was going to have with Ambassador Esenbal.

[132] Memo dated January 16, 1971 from Frank Cash, Country Director, Turkish Affairs, to the State Department, Box 2639 NARA.

In January 1971, a U.S. Army police sergeant named Jimmy R. Finley who was on guard duty at a U.S. installation was kidnapped, but then released shortly thereafter.[133]

The most spectacular event occurred on March 4, 1971 when the THKO kidnapped four air force personnel and demanded $400,000 ransom from the U.S. government. The four men were all radar technicians and had apparently been kidnapped as they were driving in a military car from a radar base to their billets in Ankara.[134] Two days later the Turkish troops stormed the Middle East Technical University as part of a nation-wide search for the kidnapped G.I.s. Two people were killed and more than a dozen were wounded as students fired back at the soldiers and allegedly threw dynamite sticks.[135] The following day, *The New York Times* reported that five people had been killed in rioting related to the kidnappings. Although the THKO had threatened to kill the Americans if the ransom was not paid, leftist sources believed that they would not do so because they had already accomplished their goal of disgracing the Demirel government as well as the police forces, demonstrating their own strength, and inciting further outbursts of violence.[136] Having found no trace of the kidnappers on the campus of METU, the Turkish authorities began to shift their search to the upper-class residential districts of Ankara. Thousands of militiamen were scouring the country and the United States had even provided the Turkish government with reconnaissance airplanes and jeeps to use in the search which were normally used for opium control. Although the air force technicians were subsequently released unharmed, this event is often cited as having triggered the military intervention on March 12, one week after the kidnapping. The fact that the kidnapping could happen at all was intolerable for the ranking generals (and perhaps also for Washington) who wanted to show that in the last resort, they were still in charge.[137] As soon as the military had issued the ultimatum, the Demirel government resigned.

The military intervention and imposition of martial law did not deter those who were engaged in armed struggle. In May 1971, the Israeli consul general Ephraim Elrom was kidnapped and killed, which led to an even larger and more thorough manhunt than after the kidnapping of the four airmen. In addition to rounding up the terrorists, a number of left-leaning journalists, professors, trade unionists, and politicians were also imprisoned. Later in the year,

---

[133] In an interview Aydin Çubukçu explained that the sergeant was abducted because the kidnappers hoped that they could steal weapons from the military, but had no intention to harm Finley. When it was discovered that both the kidnappers and Finley, who was an African-American, sympathized with the Black Panthers, Finely was given money and then released.

[134] The Air Force men were identified as Sgt. James J. Sexton, Airman 1st Cl. Larry J. Heavner, Richard Caraszi, and James M. Gholson. "Four US Airmen Kidnapped by Extremists in Turkey," *The New York Times*, March 5, 1971.

[135] "Two Die in Ankara in Hunt for G.I.s," *The New York Times*, March 6, 1971.

[136] "Turkey continues Hunt for Airmen; Rioting Subsides," *The New York Times*, March 7, 1971.

[137] Harris 1972: 145.

Deniz Gezmiş and five other convicts escaped from an Istanbul military prison, which was most likely only possible with the assistance of the prison guards. It appeared that the left had achieved its goal of instigating divisions even within the repressive organs of the state apparatus. During this time, suspicions ran high among the armed forces, as some feared that the leftist officers may take advantage of the coup situation, whereas others feared that the right-wing officers may do the same. According to a "Martial Law Command Briefing on Terrorism in Turkey" at some point during the martial law regime, 704 people were being held in Ankara martial law prisons, of which 347 were military persons.[138]

The strategy of kidnapping NATO personnel continued under military rule. In response, the state escalated its repression of the armed activists. On March 30, 1972, special forces raided the village of Kızıldere where Mahir Çayan and other members of the THKO were holding three military personnel hostage who they had earlier kidnapped.[139] The military personnel were technicians, two of them British citizens and one of them a Canadian citizen. During the shoot-out, ten people were killed, including Mahir Çayan and other members of the THKO as well as the three hostages.[140] The operation was apparently planned by the Special Warfare Department in conjunction with the CIA and the Turkish National Intelligence Organization (MIT).[141] Kızıldere represents another watershed event in Turkish history, comparable to the Bloody Sunday events in 1969; the significance of the massacre was immediately recognized by the U.S. Consulate who referred to it as a "Götterdämmerung."[142] Civilian rule was not reestablished until September 26, 1973, after twenty-eight months of martial law. According to Georg S. Harris: "These moves, although designed in the first instance to combat anarchy and terrorism, nonetheless served efficiently to silence many of America's most virulent critics."[143] Despite the spectacular nature of the kidnappings and subsequent events, which may have indeed been partially responsible for triggering the military intervention in 1971, the activities of the armed guerrillas did not disrupt the internal operations of the American military presence. It is perhaps ironic that those groups who saw themselves as being on the far left, and whose ideology was perhaps the most militant, did not in fact have the most significant impact on the U.S. military. In the next chapter, I show how a similar dynamic can also be observed in West Germany.

[138] Memo from the Embassy in Ankara to the State Department, February 1, 1973.
[139] Memo from the Embassy in Ankara to the State Department, March 31, 1972.
[140] "Kızıldere Katliamı Dosyasını Açın," March 30, 2006 (www.bianet.org). Saffet Alp was a member of the Revolutionary Proletarian Organization and also belonged to the Turkish Air Force. He was among those who were killed in Kızıldere. Because some have claimed that he was first kept alive and then later executed, attempts are being made to reopen the case.
[141] Ganser 2005: 232.
[142] Memo from the U.S. Consulate in Istanbul to the Secretary of State in Washington, DC, April 1972
[143] Harris 1972: 146.

Consequences of Social Unrest in the 1960s

By way of returning to my original question about the causes and consequences of social movements against the American military presence, I would like to refer to a report entitled "U.S. Overseas Bases: Problems of Projecting American Military Power Abroad," which was written by Admiral Thomas H. Moorer and Alvin J. Cottrell. Admiral Moorer had served both as Chief of Naval Operations from 1967 to 1970 and then as the Chairman of the Joint Chiefs of Staff from 1970 to 1974. The fact that someone who had been head of all the armed forces, holding the most powerful position in the U.S. military, would publish a report about the protests in Istanbul against the Navy is telling. The following passage addresses the impact of the Sixth Fleet protests:

The use of ports for "showing the flag" has been of no practical value in Turkey since 1969. If it is not feasible to make naval visits to Istanbul and Izmir for "flag showing" then it is impossible to implement a useful policy of diplomatic visits in Turkey at all. The only ports the United States has been safely able to visit with any frequency since 1969 are in remote and sparsely populated areas where the visits have little or no political impact on the population.[144]

The land-based forces were also impacted by the social unrest. In a letter from H. G. Torbert, Jr. Acting Assistant Secretary for Congressional Relations to Senator John Stennis on August 25, 1969 the issue of the changing U.S. profile in Turkey is addressed:

The concerned agencies of the U.S. government are currently reducing the size and urban concentration of this presence, but it is very important ... that the defense cooperation between the two countries continue.... [I]t would be unrealistic to hold out the hope that only goodwill will prevail in the future. The reductions and relocations of our military elements will, however, reduce their vulnerability to incidents.[145]

Finally, the Americans were unable to achieve a change in the Turkish constitution, which would have made strikes of Turkish employees at U.S. facilities illegal. It was not until after the coup in 1980 that this was possible.[146]

By the end of the protest cycle that I am studying (1966–1971), the navy was no longer able to make port calls, the United States had drastically decreased the number of troops stationed in Turkey, those personnel that remained were increasingly segregated from Turkish society, the attempt to criminalize labor unrest at U.S. bases had failed, and forty of the fifty-four bilateral agreements had been modified to meet Ankara's demands. Many of these issues, including the use of Turkish territory, airspace, and ports returned during the dispute over the creation of a northern front for the 2003 intervention in Iraq. Furthermore,

---

[144] Cottrell and Moorer: 1977.
[145] Letter from H. G. Torbert, Jr. Acting Assistant Secretary for Congressional Relations to Senator John Stennis on August 25, 1969.
[146] Interview with Çetin Soyak in Ankara on June 29, 2007.

there was a widespread sense that the American military presence was not creating or providing security for the Turkish people, but rather undermining it. In this regard, the military had lost its former appearance of benevolence and had been transformed from legitimate protection into what I have termed pernicious protection. Although the social unrest had not resulted in the complete withdrawal of U.S. forces, it had created social limits to what the U.S. military commanders could do within the boundaries of Turkish geography.

The current project is limited to investigating the impact of social unrest on the U.S. military presence. However, much more could obviously be said about the impact of the unrest on Turkish society. Nur Bilge Criss has argued that anti-Americanism was at the heart of the struggle between leftist and rightist forces, which led to "the most prolonged era of terrorist violence in modern Turkish history."[147] The introduction of religion into politics is another legacy of this period. In an effort to challenge leftist hegemony within the universities, an amendment was made to the National Education Basic Law that allowed graduates of imam-lycees (religious high schools) to enter the universities.[148] Therefore, the rise of political Islam and the rise of guerrilla warfare can both be traced to struggles surrounding the U.S. presence in Turkey and its bearing on the issue of Turkish sovereignty.

THE U.S. PRESENCE AFTER THE PROTEST CYCLE

1970s: The Arms Embargo and the Suspension of the Treaty

The U.S. response to the resistance of the 1960s was radical by most standards: the number of U.S. personnel in Turkey was reduced from 30,000 to approximately 16,000 in 1970 (four years before the Cyprus crisis in 1974) and to around 10,000 by 1979,[149] long before anyone could imagine the end of the bloc confrontation. By simply eliminating the bulk of the U.S. presence, it was assumed that bilateral relations could continue more or less peacefully. However, it was not long before this assumption was soon proven to be mistaken.

The defining event of the 1970s in terms of U.S.-Turkish relations was undoubtedly the Turkish invasion of Cyprus in 1974 and the subsequent arms embargo declared by the U.S. Congress on the grounds that U.S.-supplied weapons were unlawfully used in the invasion. According to the Turkish side, if Washington was no longer providing arms and aid, Turkey no longer had any reason to provide space for U.S. facilities, and after five months of the embargo, Ankara took the unprecedented step of ordering the suspension of all activities at U.S. bases. This meant that out of the twenty-seven bases in the

[147] Criss 2002.
[148] Ibid.
[149] Karasapan 1989: 7.

country, only one – İncirlik – remained in operation. Here it should be noted that the Turkish side knew very well that bases in Turkey could be divided into two different categories: those that benefited both Turkey and the United States or NATO in general, and those which primarily served U.S. interests. Examples of the latter included monitoring stations in Karamursel, Sinop, Diyarbakir, and Ankara. In fact, these bases may have even increased the risk to Turkey, although they were enormously important to Washington because 25 percent of NATO's hard intelligence concerning the USSR was gathered from sites in Turkey.[150] It is clear, therefore, that the closure of these sites had a serious impact on U.S. monitoring activities and that, despite NATO, Turkish and U.S. interests did not always converge.

The closure of the U.S. military bases is perhaps the most dramatic illustration of how the U.S. protection regime had transformed from one of legitimate protection into a form of pernicious protection. If the Turkish people had felt truly protected by the U.S. bases, it is unlikely that they would have been ordered to suspend their activities.

Because of these diverging interests, it should come as no surprise that their perceptions of the same event were very different as well. The headlines of articles in major American newspapers highlighted the sense of vulnerability on the part of the U.S. military authorities: "Americans in Turkey Encountering Signs of Strain,"[151] "Turks Move into U.S. Bases: Extent of Takeover Unclear,"[152] and – perhaps most dramatically – "Turkey holds ax over U.S. bases."[153] Apparently this Turkish "take over" of what the Americans perceived as "their" bases caused more than one military commander to be concerned. The social unrest in Turkey had highlighted just how dependent the U.S. military presence was on both the local infrastructure as well the labor power of base workers.

Relations between Ankara and Washington deteriorated rapidly and then came to a standstill. Five years after the Cyprus crisis was triggered, a report by the CSIA working group at Harvard observed that "the southern flank remains in crisis" and that the issue was not receiving the attention it merited. Furthermore, it warned that alarmism about the Soviet threat was unlikely to lead to rapprochement because Turkey and Greece now viewed each other as a more serious threat to their security than Moscow.[154]

Although the arms embargo was lifted in August 1978, a new defense agreement was not signed until the 1980 Defense and Economic Cooperation Agreement (DECA). Several points are important to make in this regard. First, the embargo contributed to hardening public opinion against the United States, which essentially gave credence to the arguments of the left while weakening

[150] Kuniholm 1985: 229.
[151] *The New York Times*, February 13, 1975.
[152] *The Washington Post*, July 27, 1975.
[153] *Christian Science Monitor*, August 11, 1975.
[154] CSIA 1978/79: 176.

the pro-NATO position.[155] Second, the DECA was never approved by the Turkish parliament but was signed by the National Security Council (NSC) under Kenan Evren. Finally, and perhaps most importantly, this meant that for half a decade, from 1975 until 1980, there was no legal basis for the U.S. presence in Turkey. This was one of the most dramatic and sudden loss of facilities within NATO during the Cold War.

1980s: Martial Law and Reducing the Visibility of the U.S. Military

The Soviet invasion of Afghanistan and the Iranian Revolution in 1979 once again increased Turkey's strategic importance in the eyes of Washington. The fall of the Shah meant that the United States had lost not only one of its key allies in the region, but also its military presence in Iran. The failure of the hostage crisis can at least in part be attributed to the fact that the infrastructure in Turkey was still unavailable, and that the United States had to operate from bases in Egypt.[156] According to a U.S. Congressional Research Service Report published in early 1980, Turkey's crisis is "the most immediate threat to US interests" and Congress should "deal with this urgently."[157]

On September 3, 1980, General Tahsin Şahinkaya, the Commander of the Turkish Air Force, paid a visit to Washington for high-level talks. A little more than a week later, on September 12, the High Command intervened for the third time since 1960 and declared a state of martial law. General Şahinkaya became a member of the military junta, while General Kenan Evren was made chairman of the National Security Council that ruled the country until 1983. The coup had been preceded by several years of violent unrest that had escalated to a level approaching civil war. By some accounts, more than 5,000 political assassinations took place during the late 1970s; street violence also included attacks against U.S. personnel and American institutions in Turkey. Between April 1979 and December 1980, nine Americans were killed; the Turkish-American Association and a civil engineering facility were bombed; bombs exploded at a hotel in Izmir leased by the U.S. military, at offices of the local Turkish-American Association, and at NATO's Rod and Gun club; and gasoline bombs were found under American-owned cars.

Because the number of U.S. troops in Turkey had already been drastically reduced, the immediate response was now to reduce the visibility of those who remained. Because it had been determined that Americans were under the greatest threat when they were going to and from work, U.S. personnel were no longer required to wear uniforms and some were allowed to start work an hour later, so that they would not be on the streets before sunrise.[158] Finally, in addition to reducing the number of Americans in Turkey as well as the visibility

---

[155] Thompson 1996: 111.
[156] Thompson 1996: 109.
[157] *Christian Science Monitor*, March 11, 1980.
[158] *The New York Times*, December 9, 1980.

of those who were still there, the number of facilities was also cut. Of the previous twenty-seven military installations, the Department of Defense decided that it no longer needed thirteen of them.[159]

This chapter cannot provide an adequate account of the impact of the 1980 coup on social movements or Turkish society more generally.[160] It was, however, the most violent of the three military coups that had occurred in the two decades since 1960, and had the most decimating affect on civil society.[161] The parliament was shut down, the constitution was suspended, and political parties were banned. It is estimated that 23,667 associations were shut down and that 650,000 people were arrested.[162] U.S. officials claimed that they were not informed about the coup, but some analysts consider this unlikely as the United States "effectively operates Turkey's military communications system."[163] Afterward, however, U.S. endorsement of the coup could be gleaned from the fact that NATO's commander General Bernard Rogers allegedly visited Ankara four times during the month following the intervention. And in November, General David Jones, Chairman of the Joint Chiefs of Staff, visited Turkey. During the same month, the World Bank announced a loan of $87 million.[164]

It is safe to say, however, that during the 1960s and 1970s anti-imperialism was perhaps what best united the various factions on the highly diverse left spectrum and that protest against the U.S. presence featured prominently within this agenda. Nevertheless, after September 12, 1980, the left essentially found itself outlawed and social unrest came to a standstill.[165] Those who managed to escape, did, and the Turkish diaspora in Western Europe was enlarged to include leftist intellectuals, activists, and trade unionists. According to Çağlar Keyder, the primary institutional legacy of three years of military rule (1980–1983) was the expansion of the powers of the National Security Council (NSC), forming a "parallel government" as well as the expansion of the State Security Courts, forming a "parallel legal system" that had jurisdiction over crimes against the state. These "crimes" were defined as ranging from separatist propaganda, to the public display of religious beliefs, to the singing of a song in the Kurdish language. "Virtually everything, from foreign and military policy to the structure of civil and political rights, from secondary-school curricula to energy policy, was eventually decided in the monthly meetings of the NSC, invariably along the lines formulated by its secretariat."[166] Beginning

---

[159] Gonlubol 1975.

[160] For a detailed account of the coup, see Mehmet Ali Birand's *The Generals' Coup in Turkey: An Inside Story of 12 September 1980.*

[161] Karasapan 1989: 8; "Turkey: Aging Generals Face Trial for Violent 1980 Military Coup," *The New York Times*, April 4, 2012.

[162] See "12 Eylül darbesinin bilançosu" www.cnnturk.com/2010/yasam/diger/09/06/12.eylul.darbe-sinin.bilancosu/588453.0/.

[163] "The Coup," Jim Paul, MERIP, no 93, 1981: 3.

[164] Ibid: 4.

[165] Lipovsky 1992.

[166] Keyder 2004: 67.

with Özal in 1983, civilian governments were entrusted with the management of the debt and economic policy. One of the first structural adjustment packages that the IMF implemented was in Turkey. The fact that Ankara received money because of its "special geopolitical position" led the *Financial Times* to comment that Turkey was an "expensive ally."[167] Under these circumstances, the U.S. and Turkish militaries were able to renew their strategic partnership with the 1980 DECA agreement, the primary purpose of which was to maintain access to the facilities that the United States had temporarily "lost."

From the perspective of the governing elites, the decade of the 1980s represented a return to normalcy as relations were gradually mended. A number of publications argued that Turkey had become a "neglected" or "forgotten" ally because of the tribulations of the previous years and that it was necessary to reevaluate Turkey's importance for the United States. Such publications included: *Turkey, the Alliance, and the Middle East* by Paul B. Henze,[168] *Turkey in Transition: the West's neglected Ally* by Kenneth Mackenzie, *Turkey and the West* by David Bachard, *Turkey: Coping with Crisis* by George S. Harris, and Dankwart Rustow's *Turkey: America's Forgotten Ally.*[169]

With the Cold War resurging under President Ronald Reagan and academicians advocating a reevaluation of Turkey's importance, U.S. defense officials dusted off war plans from the 1950s, the "golden age" of the U.S. presence in Turkey. Former University of Chicago professor Albert Wohlstetter played an important role in reviving these plans and several of his students, including Paul Wolfowitz and Richard Perle, would serve in important positions under Reagan. Perle, in particular, was a driving force behind implementing Wohlstetter's plans for Turkey. During his seven-year term at the Pentagon, Perle supervised U.S. aid to Turkey and was a staunch supporter of Ankara during Congressional hearings. After he resigned from his post, he created a lobbying firm, International Advisors, which Douglas Feith and Michael McNamara later joined. Perle also played a major role in establishing the Turkish-Israeli defense alliance.[170] It was during this time that a series of bilateral agreements was signed, some in the context of NATO, the most important consequences of which were the militarization of the Kurdish region of eastern Turkey. More than thirteen air bases were either built, rebuilt, or modernized.[171] Turkey received $2.5 billion in aid over the next five years and a 1982 Air Bases Operating Agreement allowed the United States to preposition equipment. Finally, Turkey repositioned its most capable forces to the southeast: the Headquarters of the Second Army Corps moved eastward, from Konya to Malatya, whereas Konya became a base for AWACS planes. The mid-1980s also witnessed the reorganization of the Turkish

---

[167] Cited in Thompson 1996: 111.
[168] Henze was a former member of Carter's National Security Council and served as CIA station chief in Turkey.
[169] Henze 1982; Mackenzie 1984; Bachard 1985; Harris 1985; Rustow 1987.
[170] Evriviades 1998.
[171] Marios L. Evriviades, "The Evolving Role of Turkey in U.S. Contingency Planning and Soviet Reaction," paper presented to the John F. Kennedy School of Government in 1984.

defense industry ordered by Act 3238. In 1985, the National Security Council drew up the National Military Strategic Concept which outlined a ten-year modernization plan for the armed forces. The result was to create a "systematic process for linking Turkish national strategy to defense acquisition."[172] In sum, despite the loss of U.S. credibility during the previous decade, after the 1980 coup, elites were able to reestablish a functional protection regime, albeit one that was still widely regarded as pernicious or problematic. Although the legitimacy of the protection regime could not easily be restored once it had been lost, reestablishing its functionality was not only possible, but deemed necessary in the face of a still existent Soviet threat.

1990s: The Gulf War and the No-Fly Zone

If during the 1960s and 1970s, the focal points of social unrest were primarily in the urban centers of western Turkey (such as Istanbul, Ankara, and Izmir), as well as other smaller cities, from the mid-1980s onward, the geographical center of resistance had switched to the southeast and the countryside. The 1980 coup and three years of martial law had put a halt to social movements and many members of leftist organizations found themselves in prison or in exile. Having effectively crushed social unrest in the West, the Turkish state turned its attention toward the East. In 1984, the Kurdish Workers' Party (PKK) began its armed uprising and maintained strongholds in Diyarbakir, Tuncele, Antep, Mardin, and a number of other cities spread across eastern Anatolia.

The United States, however, was more interested in Iraq and the situation of the Kurds there than in the Kurdish question in Turkey.[173] The details of the Iraqi invasion of Kuwait and the subsequent U.S.-led war against Iraq are well documented. Less well known may be the fact that former President Turgut Özal not only enthusiastically supported U.S. policies, but even wanted to open a northern front against Iraq and send Turkish troops to join the coalition forces in the Gulf. General Necip Torumtay, then Chief of the General Staff, explains in his memoirs that this was one of the main reasons for his resignation on December 3, 1990.[174] Clearly there was disagreement within the establishment over the issue of the Gulf War. Despite his objections, Torumtay gave in because "behind Özal stood the shadow of General Evren," who was perhaps the most powerful figure in the pro-U.S. camp within the military.[175]

Because Özal encountered resistance not only within the opposition parties, but also within his own Motherland Party and the military, Turkey's policy of allowing the coalition to use the air bases in İncirlik, Diyarbakir, Malatya, and

---

[172] Hickok 2000.
[173] For a number of years, Ankara had been pressing various American administrations to officially declare the PKK a terrorist organization; however, Washington did not do so until 1994, ten years after the insurgency had begun.
[174] Hale 1999: 102; "US Prods Ankara on Iraq," *Turkish Daily News*, December 23, 2002.
[175] Hickok 2000.

Batman, closing off the Kerkuk-Yumurtalik pipeline and the ending of all regular trade with Iraq was a compromise between Özal's ambitions and reluctance
within other branches of the government and public opinion.[176]

The ramifications of this "reluctance" are well illuminated with an anecdote
by Morton Abramowitz, who was the U.S. ambassador to Turkey from 1989
to 1991. He recalled how he was awoken in his private residence in Ankara by
a phone call at 2 a.m. on January 16, 1990, from Dick Clarke, then Assistant
Secretary of State for Political-Military Affairs. Clarke told Ambramowitz that
the B-52s had just left Spain for Iraq and that they needed overflight rights over
Turkey immediately. He said the jets would reach Turkish airspace in about
six hours, or around 8 a.m. As Abramowitz could not possibly formally secure
overflight rights in the middle of the night, he asked if the planes could fly over
Syria instead. The answer was "no." Abramowitz writes:

> I was never thrilled, as an ambassador, to call the country's leader in the middle of the
> night. The president of Turkey actually lived across the street, but I picked up the phone
> and called.... I told the president the United States needed overflight rights over his
> country within the next five hours. His only question was, "Can they fly over Syria?" I
> simply said no, we didn't want to ask and they wouldn't agree. Özal said he would call
> back. He did, a short while later, and asked if the planes could fly close to the Syrian
> border so the overflight could be plausibly denied in Turkey. I said yes and would tell
> Washington immediately.[177]

As this anecdote demonstrates, at the end of the twentieth century and after
nearly fifty years of bilateral security relations between the United States and
Turkey, some agreements were still made on the telephone, in the middle of the
night, on an ad-hoc basis.

Political scientists who either point to the anarchic nature of the interstate
system where anything is possible or to the realm of Realpolitik marked by
scarcity and constraint, may not raise an eyebrow at such incidents, but the
Turkish populace, who has more than once contended the decisions of the elite,
did. In fact, according to Abramowitz, the use of bases during the Gulf War
was most likely the single most controversial issue in Turkey at that time.[178]
Beginning in April 1991, six weeks after the end of Operation Desert Storm,
İncirlik was used to patrol the no-fly zone over Iraq and hence the basing issue
remained "the most important specific factor in U.S. policymaking toward
Turkey for much of the past decade."[179] The use of İncirlik for the northern
no-fly zone, as it turned out, was not much less controversial than the use of
İncirlik during the bombing campaign. This is apparent in the fact that what
the Americans referred to as Operation Provide Comfort was translated into

[176] Sever 2002: 27.
[177] Abramowitz 2000: 160.
[178] Ibid: 155.
[179] Ibid: 157.

the Turkish language as Operation Poised Hammer.[180] Despite the unpopularity of the no-fly zone and the Gulf War, this wide-spread sentiment did not result in any significant anti-war movements in Turkey.[181]

It is difficult to overemphasize the importance of the İncirlik Air Base for U.S. power projection in the Middle East during the end of the twentieth century; for more than a decade, the entire Iraq policy of the United States hinged on İncirlik. When one considers that Iraq was either at or near the top of the foreign policy agenda for three subsequent administrations, one could argue that İncirlik was perhaps the single most important overseas base in the world.[182] The İncirlik Air Base is undoubtedly one of the reasons that Turkey was elevated to the role of a "strategic partner" of the United States during former President Clinton's visit to Turkey in November 1999. After the end of the Cold War, a number of foreign policy experts across the political spectrum emphasized Turkey's increasing importance: Brzezinski referred to Turkey alongside Russia as the two most important Eurasian powers, Richard Perle described Turkey as a "front-line state" whereas others point out that it is a "European, Balkan, Middle Eastern, Near Eastern, Caucasian, Mediterranean, Aegean and Black Sea power."[183] According to a Pentagon report from 1995:

Turkey in particular is now at the crossroads of almost every issue of importance to the United States on the Eurasian continent – including NATO, the Balkans, the Aegean, Iraqi sanctions, relations with the NIS, peace in the Middle East, and transit routes for Central Asian oil and gas.[184]

Turkey was no longer neglected or forgotten, but seemingly at the center of everything.[185] Despite the fact that this would appear to give Turkey some degree of bargaining power with Washington, and despite the unpopularity of the no-fly zone among the Turkish population, there was little that Ankara could do to change the status quo. When in opposition, political parties have

---

[180] In August 1992, the southern no-fly zone was imposed to prevent Iraqi flights south of the 32nd parallel and were intended to protect the Shiite population; in 1996 Clinton expanded the southern no-fly zone to the 33rd parallel. *The Washington Post*, June 16, 2000.

[181] There are perhaps two reasons for this: (1) the war was sanctioned by the United Nations and hence legitimate even if unpopular; and (2) the Turkish civil society was still recovering from the aftermath of the military regime, which had ended only seven years before the war began in January of 1991.

[182] The southern no-fly zone was enforced by planes stationed at bases in Kuwait, Saudi Arabia, and aboard U.S. aircraft carriers in the Gulf; hence, it was not as dependent on one single air base as the northern no-fly zone.

[183] Simon V. Mayall, Turkey: Thwarted Ambition, McNair Paper No. 56, Washington, DC, Institute of National Strategic Studies, National Defense University, January 1997: 1.

[184] Office of International Security Affairs, "United States Security Strategy for Europe and NATO," edited by Department of Defense: DoD, Washington DC, 1995.

[185] Some have discerned an Ankara-Tel-Aviv-Cairo axis that by the mid-1990s had replaced the failed CENTO alliance. In 1994–1995, 40% of all F-16 exports went to Turkey, Israel, and Egypt. Thompson 1996: 107.

often opposed the no-fly zone, but then supported it when in office. The reversal made by the Welfare Party after it came to office in June 1996 is just one example.[186] General Kenan Evren, the leader of the 1980 military coup and later President between 1982 and 1989, summed up the situation : "Whoever is in power cannot cancel Poised Hammer; because we have been too dependent on the United States, both economically and politically."[187] As demonstrated in Chapter 5, this status quo was changed dramatically in 2003.

CONCLUSION

Over the course of more than half a century, the nature and size of the U.S. presence in Turkey has changed considerably, with the end of the Cold War generally regarded as representing the single most important turning point in the international system as a whole, and in terms of U.S.-Turkish relations in particular. However, there were substantial changes during the Cold War as well. In fact, if the most important classification of overseas troops is whether or not they are welcome by the host country, as a report by the Heritage Foundation has recently suggested, then it would be safe to say that whereas U.S. personnel were welcome in Turkey during the late 1940s, it would be hard to make the same claim for the 1960s.[188] A comparison of how the USS Missouri was welcomed to the harbor of Istanbul in 1946 and how the Sixth Fleet was greeted in 1968 should be a vivid demonstration of how the times had changed.

How can the demands put forth in the parliament, in the streets, and on the bases themselves be summarized? The demands which were raised by members of the Turkish Labor Party, the student demonstrators, and the base workers were of course at times contradictory, with some wanting to increase Turkish control over the U.S. presence and others wanting to end the U.S. presence altogether. They did not, however, contradict each other in the simple fact that they expressed some form of popular sovereignty. It was this layer of social forces below that of the state that had to be reckoned with in order to pacify the turbulent relationship and maintain the alliance. In the next chapter, I analyze how similar dynamics played out in a different historical context in West Germany.

[186] Hale 1999: 102.
[187] Cited in Candar 2000: 139.
[188] Kane 2004.

3

Social Unrest and the American Military
Presence in Germany during the Cold War

INTRODUCTION

During the Cold War, the Federal Republic of Germany hosted more foreign troops, military bases, and weapons stockpiles than any other country in the world. At any given point during the Cold War, nearly three-fourths of the European-based forces were stationed there, or approximately 250,000 personnel, not including family members. For the U.S. Army and Air Force, it is still today the single most important host nation in Europe; with 70,000 troops, it is one of the largest overseas deployments in the world. Depending on how the facilities were counted, the Federal Republic was home to between 600 and 900 U.S. military facilities during the Cold War, including the headquarters of the European Command (EUCOM), the only U.S. Command Headquarters that is located outside of the United States.[1] In addition to military facilities, there was an extensive civilian support infrastructure including schools, churches, shopping malls, and golf courses (see Table 1 in the appendix for a full list). When these civilian and military facilities are added together, the total number of U.S. facilities in West Germany during the Cold War was approximately 1,500.[2]

Although NATO strategies changed over time, the geographical location of Germany ensured that it remained the Central Front and hence potential battleground between NATO and the Warsaw Pact.[3] For this reason, German

---

[1] An atlas published in 1986 by the Green Party, which gave detailed information about the location and purpose of every military facility operated by the United States listed 902 bases, whereas a 1989 study by the Stockholm International Peace Research Institute, which combined closely associated facilities, listed 625. See Luber 1986. In 2008 a new regional command was created: the United States Africa Command (AFRICOM) which currently has its headquarters in Stuttgart, Germany, where EUCOM is also located.

[2] Bebermeyer 1989: 56.

[3] I will use the terms West Germany, Germany, and FRG interchangeably, although it is clear that I am referring to the Western part of pre-unification Germany during the Cold War.

territory and airspace, as well as the entirety of its civilian infrastructure, served as a vast training ground for a diverse array of military maneuvers and practice exercises for more than four decades.

An entire genre of academic literature has shown how postwar Germany was shaped by the policies of the United States and other allies; it was denazified, decartelized, democratized, and (temporarily) demilitarized. Strangely, few have pointed out that it was also shaped by the *physical presence* of foreign militaries, in particular the U.S. Army and Air Force. In this chapter, I hope to highlight that the rise of protest movements in the 1970s and 1980s made it necessary for U.S. military authorities to respond to the changing political climate and that the U.S. presence was, in turn, shaped by the opposition it encountered. By analyzing the activities of the Red Army Faction (RAF), the Revolutionary Cells (RZ), the non-violent peace movement, labor unrest of base employees, and the Greens, I hope to illustrate the causes and consequences of anti-base unrest in West Germany.

Most of the literature on the U.S. presence in Germany focuses on the early occupation period from 1945 to 1955.[4] Of course, the U.S. presence was not a static entity, but underwent dramatic changes since the original occupation; the 1970s–1980s was a particularly tumultuous period. Several decades of nearly seamless cooperation between the two allies had led not only theorists of democratization but also basing strategists to consider German-U.S. cooperation as a major success story of the postwar period. The NATO double-track decision in 1979, however, led to a serious deterioration of bilateral relations and is usually cited as having triggered the largest social movement in postwar German history. This decision not only gave rise to an upsurge of protest activity, but also led to the fall from power of the Social Democratic Party (SPD) in 1982 and helped to catapult the Green Party from local politics into the Bundestag in 1983. The crisis was particularly severe because it was a crisis "from below," or an all-encompassing societal upheaval that revealed profound differences in how Germans and Americans perceived the security landscape. Perhaps unsurprisingly, the large peace rallies have attracted the most attention within the social movement literature because these were not only the most visible but for many perhaps also the most spectacular protest events. Not only were they the largest demonstrations during the half-century of the Cold War, but many of the speakers were prominent personalities such as former Chancellor Willy Brandt or the winner of the Nobel Peace Prize, Heinrich Böll. Instead of focusing on the mass demonstrations held in city centers, I describe those protest events which directly targeted U.S. military facilities.

These oppositional groups ranged from grassroots citizen initiatives that were located next to small facilities in rural areas and were committed to nonviolence, to internationally operating terrorist networks that carried out attacks against the headquarters of the U.S. Army in Europe and other major

---

[4] Examples of studies which concentrate on the early years of the U.S. presence include: Glaser 1946, Davidson 1959, Gimbel 1968, Höhn 2002, Goedde 2003, Grossmann 2007, Zepp 2007.

installations. Clearly, these groups did not form a coherent whole in the sense that they represented a unified movement. On the contrary, activists often came from opposite ends of the political spectrum and saw one another's activities as detrimental to their own. Scholars have tended to study these groups separately. However, from the perspective of the U.S. Army Liaison Officers who were reporting back to Washington on "host nation activities," all of this represented a challenge to their authority and to the American military presence more generally.[5] By bringing together these diverse strands of political turmoil, I am by no means intending to equate them. Nor do I agree that they all represent a primarily cultural expression of some diffuse form of anti-Americanism, as others have charged.[6] Instead, despite varying ideologies and tactics, I believe that these different forms of contentious politics all took issue with what they saw as the increasingly pernicious nature of the U.S. protection regime. For some groups, the harm (whether potential or real) caused by U.S. military activities was what motivated them. For others, it was the fact that they felt disenfranchised as they had little or no say in U.S. military operations on German territory. The same military bases that were potentially able to execute intentional, concrete harm towards the Soviet Union, were supposed to represent protection to NATO allies. And yet, in the eyes of many, the U.S. bases caused collateral harm, whether in the form of environmental degradation, military plane crashes, or the stationing of nuclear missiles, to name just a few examples. The fact that the collateral harm was being caused by a foreign military presence that operated with considerable autonomy only exacerbated the problem, by disenfranchising the people they intended to protect.

As with my discussion of Turkey in Chapter 2, I have divided the large spectrum of anti-base unrest in Germany into four categories: nonviolent civil disobedience, protest in the parliament, labor strikes, and violent attacks by armed groups. Before coming to the protest cycle of the 1970s and 1980s, it is necessary to provide a very brief background to the early postwar period.

BEFORE THE PROTEST CYCLE

Establishing the U.S. Presence in Germany

Under the leadership of Supreme Allied Commander General Dwight D. Eisenhower, more than 3 million G.I.s had fought in the European theater;

---

[5] The system of liaison officers in Germany was a relic of the occupation. Each German state within the American sector was assigned one liaison officer and, depending on the size of the Bundesland and the U.S. presence there, one or two regional officers as well. The liaison officers would write biweekly reports back to the Deputy Chief of Staff of Host Nation Activities (DCSHNA) which included commentary on political developments and protest activity. Because Turkey did not have a similar system of liaison officers, such archival material is not available for Turkey.

[6] Diner 1996, Müller 1986, and Pohrt 1982, among others, all criticize the anti-Americanism of the peace movement. For a discussion of anti-Americanism in comparative perspective see Katzenstein and Keohane 2007.

and on V-E Day, 1.6 million G.I.s stood on German territory.[7] The transition from wartime mobilization to an enduring overseas presence did not happen easily. The overseas force structure actually contracted after 1945: the number of men and women in the U.S. armed forces shrank from 12 million in June 1945 to 1.5 million in June 1947.[8] Although Secretary of the Navy James Forrestal and Secretary of War Robert J. Patterson believed that demobilization would endanger the U.S. strategic position in the world, the Truman White House could not overcome the overpowering isolationism within Congress, public opinion, and last but not least, also within the rank-and-file of the armed forces. Therefore, the hastily drawn-up wartime arrangements regarding hundreds of facilities were relinquished because of a lack of domestic support for an overseas commitment both within the U.S. Congress and within the rank-and-file of the U.S. military as G.I.s refused occupation duties and rioted in January 1946, impatient to be sent home.[9] On January 10, 1946, *The New York Times* reported that G.I.s were in a "mutinous mood" and had to be held back with bayonets from charging the U.S. Army headquarters in Frankfurt.

The impact of demobilization within Germany meant that the military machine, which had played an important role in defeating the Third Reich, shrank to something resembling more a police force with bad equipment and low morale than a militarily effective occupation regime.[10] These troops were at best able to maintain public safety, but were not able to act as a real counter balance to the USSR. By 1950, there were only 79,000 G.I.s in Germany.[11]

The disintegration of the anti-fascist wartime alliance and the onset of the Cold War have been well documented. As early as 1946, Churchill referred to an Iron Curtain descending over Eastern Europe, and diplomat George Kennan described the Soviet Union as determinedly expansionist in his famous Long Telegram. However, it required more than just a speech and a telegram to convince the public of the dangers lurking behind the so-called Iron Curtain. Usually, a number of events are cited as leading to the growing hostility between the United States and the USSR, these included: the dissolution of the Allied Control Commission governing Germany, the Communist takeover of Eastern Europe, the Russian acquisition of the bomb, the Chinese Communist victory, and the Berlin Blockade and subsequent Airlift. Because of its direct impact on the U.S. presence in Germany, the Berlin Crisis was certainly the most significant of these events for the current discussion.

In June 1948, the United States and Britain unexpectedly announced a currency reform for West Germany which meant that all old Reichsmarks became worthless and had to be exchanged for the Deutschmark. When the reform was

---

[7] Schraut 1993.
[8] Pollard 1985.
[9] "Bayonets Disperse G.I.s in Frankfort," *The New York Times*, January 10, 1946.
[10] Schraut 1993: 153.
[11] Baker 2004: 48.

extended to Berlin, the Russians shut off all land routes to the city to protest being excluded from negotiations concerning Berlin, which after all was in the Soviet occupation zone. Rather than interpreting this as merely a demand to be included in future negotiations, the United States saw it as a challenge to the "Free World" and responded with the eleven-month Berlin Air Lift. This event not only brought West Germany solidly into the Western Alliance, but also provided the details to create the epic narrative of legitimate American protection against a hostile Soviet Union. The "Rosinenbomber" that distributed food and even candy to West Berliners became a legendary symbol of the benevolent protection offered by the United States and its allies.[12] American G.I.s were no longer perceived as part of a foreign occupation force, but, according to some, as fatherly figures.[13]

By the time the Berlin Air Lift ended within ten months, the U.S. military had been transformed from an occupying force (*Besatzungsmacht*) into a protective force (*Schutzmacht*), the West Germans had been given a new and virtuous identity based on anti-Communism, and the USSR had become the common enemy. For a country that had begun two world wars and engineered the Holocaust, this was perhaps the best of all possible worlds.

This schematic historical outline illustrates an important conceptual point. As discussed in the Introduction, the overseas network of U.S. military bases forms a global division of violence and protection. Just as production processes occur through an international division of labor, with commodity chains linking factories, warehouses, and retail stores around the world, so too is protection and violence "produced" through the global deployment of U.S. troops and bases. This global division of violence and protection takes the form of different types of protection regimes within different countries that host U.S. bases. I have built on Tilly's conceptualization of war-making and state-making, by arguing that for a foreign military presence to qualify as offering legitimate protection toward a host nation, it must fulfill two minimum requirements: (1) it must offer protection against an outside threat which it did not create itself, and (2) it must cause no harm toward those it claims to protect. I argue that, if both of these conditions are met, the U.S. military presence constitutes legitimate protection. In the case of Germany, the threat against which the U.S. military originally claimed to offer protection was not external but internal to the host country. Having defeated the Nazi regime by the summer of 1945, the U.S. presence continued to guard against any potential

---

[12] The official military historiography literally rhapsodizes about the Berlin Air Lift with journal articles such as "Candy Bomber Delivers Happiness" by David Castellon (1998), or "The Air Force can Deliver Anything! A History of the Berlin Airlift" by Daniel F. Harrington (1998).

[13] Petra Goedde has even argued that the absence of German men, either because they had been killed on the battlefield or were still being held as prisoners of war, led U.S. troops to perceive Germany as weak and effeminate. Goedde 2003.

reemergence of German militarism, and can be considered a "precautionary protection regime." The brute force that broke the back of the Wehrmacht gave way to an occupation that had been planned years in advance and was relatively well-managed, with the German population generally acquiescing if not actively consenting to the occupation. With the onset of the Cold War, an external threat appeared on the horizon and the two basic requirements were met: the U.S. occupation was transformed into legitimate protection. This is the transformation which took place during the first five years of the U.S. presence in Germany, as briefly described previously. As demonstrated in this chapter, the U.S. protection regime in West Germany underwent another major shift during the 1970s and, in particular, the 1980s.

Containment

To return to the historical narrative, the Berlin Airlift was significant not only because of its transformative affect on the U.S. protection regime, but it also represented the onset of the larger Cold War. The U.S. response to what was perceived as an expansionist Soviet Union was the policy of containment. This policy was officially defined by the Joint Chiefs of Staff in 1950 in a document known as NSC 68.

The policy of containment included establishing a huge military presence in Europe, Asia, and the Middle East that consisted of four main elements:

1. establishing a ring of Air Force bases around the USSR and China;
2. deploying Naval Forces in the Pacific Ocean and Mediterranean Sea, where aircraft carriers could relocate along any coastline;
3. stationing ground forces of the U.S. Army mostly in Europe and South Korea; and
4. creating thousands of small intelligence-gathering installations around the perimeter of the USSR and China.[14]

Although "containment" is usually counter-posed to the "expansionist" tendencies of the USSR and described as a policy that protected the status quo by keeping the USSR in check, in reality, containment meant projecting U.S. power into the Eurasian landmass by building military outposts far away from the homeland. It was not World War II but rather the onset of the Cold War in the late 1940s and the Korean War from 1950–1953 that was the watershed: the makeshift and fragmented pre–World War II force structure was permanently transformed into a colossal baseworld with huge facilities strung across Europe and Asia. Normally, this transformation is referred to as the "troops-to-Europe" decision and, perhaps unsurprisingly considering how unpopular it was at the time, it was considered one of the most radical decisions that Truman made during his presidency. In September 1950, the decision

---

[14] Baker 2004: 49.

was made to send four divisions back to Europe to build a NATO force under the command of General Dwight D. Eisenhower.[15] However, the creation of an overseas U.S. basing network was more than just a matter of sending troops. Instead, it was about establishing a permanent foothold in a number of foreign countries and bringing European allies under U.S. command. Perhaps no country was more affected by this decision than Germany. The FRG was divided into four zones and each zone was under the occupation of one of the victorious allies: the Soviets in East Germany, the British in the North, the French in the West, and the United States in the South and Southwest. Berlin, the former capital of the Third Reich, formed a fifth zone subject to a special four-power arrangement under the auspices of the Allied Kommandatura. Berlin retained a special status throughout the Cold War as it was never part of NATO and was also divided into four occupation zones. The American sector was divided into two parts: the Western Military District (Baden, Württemberg, Hesse, Rhineland Palatinate, and the Bremen Enclave) was controlled by the Seventh Army and the Eastern Military District, consisting of Bavaria and the Berlin District, was controlled by the Third Army. Map 3.1 shows the density of the U.S. presence in the American sector of West Germany, not including Berlin.

THE OPPOSITION CYCLE BEGINS

Armed Struggle: The Red Army Faction and the Revolutionary Cells

As discussed above, and as demonstrated by Map 3.1, Germany was perhaps the most heavily militarized territory in the world. In other countries which hosted U.S. troops and bases, protest movements had already emerged in the early 1960s. The fact that this did not happen in Germany until later is related to the historical circumstances of the occupation and the fact that the United States was widely regarded as a protective power (*Schutzmacht*). Even during the height of demonstrations against the Vietnam War in the late 1960s and early 1970s, most activists directed their critique against U.S. foreign policy in Asia, but not the U.S. presence in Germany. Furthermore, these protests were mainly organized and carried out by students and youth groups, but did not expand to encompass a larger cross-section of society.

During the Vietnam War, the rank-and-file of the U.S. military had become increasingly ungovernable because of problems related to relative impoverishment, racial discrimination, drug abuse, and the politicization of enlisted soldiers because of the anti-war and Black Power movements.[16] The culture of the military was changing from one of discipline and obedience to one of insubordination – in many ways even more radical than the protest culture of

---

[15] Two divisions were already stationed in Europe, so the addition of four more divisions, made six total. Duke 1989: 59.

[16] See Höhn and Klimke 2010.

MAP 3.1. US Military Presence in West Germany during the Cold War.

civilian youth movements. For the most part, the mainstream German peace movement (which had organized Easter Marches against nuclear weapons during the 1960s) was not concerned with the internal problems of the U.S. military. There were some exceptions. A few isolated activists tried to help G.I.s escape to Scandinavia to avoid being sent to East Asia; and Franz Josef Degenhardt, a folk singer, immortalized the plight of American G.I.s in his song about PT who lived in K-town Germany. On the other hand, most members of the Rote Armee Fraktion (Red Army Faction) did not harbor much sympathy for the enlistees, but rather saw the U.S. military as a useful target for their attacks.

Similar to the THKP-C and THKO in Turkey, the Red Army Faction (RAF) and Revolutionary Cells (RZ) emerged out of the splintering of the left in the early 1970s. The RAF is considered one of the most resilient of all the European terrorist organizations, as it survived for more than twenty years despite the fact that three successive generations of leadership cadres either died or were imprisoned. Similar to their Turkish counterparts, RAF members continued their struggle from within the high-security prisons by engaging in hunger strikes and issuing statements about the changing political landscape. Despite emanating from different cultures, strikingly similar protest tactics emerged.

The RAF considered itself part of something larger, or as a "fraction" of a larger Red Army that would liberate the oppressed masses from capitalism, imperialism, and the remnants of fascism that had lived on despite the unconditional surrender of Hitler and the defeat of the Wehrmacht. According to their analysis, fascist traditions had only been briefly interrupted during the short period of denazification but then resurfaced as Germany became integrated into the Western alliance and formed a front against the Soviet Union, which had also served as the arch-enemy of Nazi Germany. The remilitarization of the Bundeswehr against the Communist East was interpreted as a continuation of a deeper German tradition. Despite the fact that the RAF borrowed its name from the Soviet Red Army, the RAF did not consider itself as defending the USSR or its ideologies, but as transporting the anti-imperialist struggles in the Third World into Europe. In addition to targeting high-ranking German politicians and business men, the RAF also targeted U.S. military installations and officials.

To achieve their goal of dismantling both imperialism and capitalism, RAF units were organized into "commandos" or "guerrillas," which were responsible for the more tactically complicated, lethal operations.[17] During the span of more than twenty years, the RAF carried out assassinations, car bombings, rocket attacks, remote-detonated roadside bombings, kidnappings, bank robberies, weapons and explosives thefts, and sniper attacks.[18] To explain its actions and ideologies to a wider audience, the RAF has issued ideological tracts, attack communiqués, trial statements, interviews, special communiqués, several issues of an underground journal, and its collected "works" in book form. *The Concept of the Urban Guerilla*, published in 1971, and *Stadtguerilla and Class Struggle* from 1972 are both by Ulrike Meinhof, who was in many ways the intellectual head of the RAF. In addition to these two publications, *Closing the Gaps of Revolutionary Theory – Build the Red Army!* by Horst Mahler is considered as having provided the ideological framework of the Red Army Faction. Furthermore, interviews with surviving RAF members such as Irmgard Möller and Thorwald Proll have been also published either in books or journals.[19]

From the beginning, the RAF had a fairly large group of sympathizers, known in German as the "*Sympathisantenkreis.*" During the early years, the RAF was even supported by prominent personalities such as Heinrich Böll, who published an article in *Der Spiegel* on January 10, 1972 that accused the German nation of engaging in a type of witch hunt against six rebellious youth. However, public opinion toward the RAF changed over time, but even after

---

[17] Although Rudi Dutsche is credited with having introduced the term "Stadtguerrilla" into the German language, it is unlikely that what he had in mind was what the RAF later became; see Kraushaar 2005.

[18] Alexander and Pluchinsky 1992: 53.

[19] Tolmein 1999 and Proll/Dubbe 2003.

the imprisonment of leading cadres, they continued to influence the German left. The RAF had connections with the *Action Directe* in France and with Palestinian groups as well. Individual RAF members were also in contact with French public figures such as Jean-Paul Sartre and Regis Debray, in whose apartment Gudrun Ensslin, Andeas Baader, and Thorwald Proll hid in Paris after escaping from Germany.[20]

A number of RAF members have become both iconic and polarizing figures, demonized by some as terrorists and celebrated by others as larger-than-life heroes and heroines. As mentioned previously, autobiographies and biographies have been written about numerous RAF members, including Ulrike Meinhof, Irmgard Möller, Thorsten Proll, Gudrun Ensslin, Andreas Baader, and many others. In contrast, until now very little has been written about the Revolutionary Cells (RZ), although for almost two decades the Office for the Protection of the Constitution (*Verfassungsschutz*) consistently considered the RZ the most dangerous terrorist group in Germany.[21] Whereas the RAF has achieved something approaching cult status, having been treated as a protagonist on the stage of German history and exposed to the spotlights of the media and the scrutiny of millions, the same cannot be said for the history of the Revolutionary Cells.

The RAF emerged in 1970–1971, around the same time as their Turkish counterparts. During this time, the RAF's main activity consisted of armed bank robberies to acquire money to fund their organization. During this early period, they were supported by many not only because it was seen as an outgrowth of the student movement but also because of the simple fact of their success: on September 29, 1970, they launched three bank raids on one day and acquired 220,000 DM in cash.[22] Perhaps encouraged by their own success, they began concocting even more spectacular feats and set their sights on none other than the commanders of the American, British, and French militaries in Berlin. In 1971, the RAF began planning to kidnap all three allied *Stadtkommandanten* (city commanders) of Berlin to create pressure to release other RAF members who were in jail.[23] Although this action was never carried out, the RAF entered into its second phase of bomb attacks in the spring of 1972.[24]

In May 1972, the RAF began its *"Mai Offensive"* or May Offense against a number of U.S. military installations. Its first target was the Headquarters of the V Corps in the IG Farben building in Frankfurt, where three powerful

---

[20] There is some uncertainty as to whether the RAF may have also had direct contacts with the Brigate Rosse in Italy. In any case, they shared similar tactics and goals, including the kidnapping of high-ranking American military officers who were stationed in Italy and Germany.
[21] One exception is the autobiography of Hans-Joachim Klein published in French: *La mort mercenaire. Témoignage d'un ancient terroriste Ouest-Allemande*. Klein 1980.
[22] Horchem 1991: 36.
[23] These included Ingrid Schubert, Monika Berberich, Brigitte Asdonk, and Irene Goergens.
[24] As reported in an interview with Irmgard Möller, see Tolmein 1999: 62–63.

bombs exploded, killing Lieutenant Colonel Paul A. Bloomquist and injuring thirteen other people. Lt. Col. Bloomquist was thus the first U.S. citizen to be killed by the Baader-Meinhof Group. Shortly afterward, RAF member Irmgard Möller drove a car full of explosives into the Campbell Barracks in Heidelberg, the headquarters of the U.S. Army in Europe. Three soldiers were killed and six others were wounded.[25] In a letter that was sent to the *Frankfurter Rundschau* newspaper on May 26, 1972, the "Commando July 15" claimed responsibility for the attack against the headquarters in Heidelberg. The reason for the attack was the announcement made by General Daniel James that the U.S. Air Force would carry out air strikes against any target it deemed worthy, whether it be north or south of the 17th parallel in Vietnam. The letter demanded an end to the attacks against Vietnam. Ulrike Meinhof declared that the attack was "our action against the strategy of annihilation of Vietnam."[26]

In terms of our conceptual framework, two points should be highlighted. First, the RAF at this time were opposed not to the unintentional collateral harm caused by the U.S. presence in Germany, but rather the intentional, concrete harm caused by the U.S. war in Vietnam. Second, the RAF clearly saw the global nature of the U.S. military presence and understood that Germany formed a central node in the overseas network of bases. Whereas other protest activities were meant to disrupt the local operations of the U.S. military, the simultaneous targeting of the headquarters in Frankfurt and Heidelberg were underlining the RAF's opposition to the global operations of the U.S. military.

In an interview with former RAF member Irmgard Möller after she was released after having spent more than twenty years in prison, she explained the rationale behind the attack in Heidelberg and the intended consequences in terms of U.S.-German relations.

Question: What was your interest in (attacking the base in Heidelberg)? Did you want to start a conflict between the US and Germany?

Möller: No, we were interested in the population, we wanted to mobilize them against the bases, against the support of the Vietnam War. We wanted to raise the stakes, so that the government would pay the price for its support of the American war in Vietnam. Then there would have been a conflict with the US.

Question: But doesn't that sound macabre from today's perspective: the sovereignty of Germany as a goal of the militant left? Ulrike Meinhof also demanded in her column in "konkret" in 1959/60 that Germany should be sovereign. But wasn't it clear that the independence of Germany would just make other crimes possible?

Möller: We didn't want Germany to be independent. This would assume, that we were interested in nation-states – but we were interested in the

---

[25] Nelson 1987: 165.
[26] Pflieger 2004.

destruction of imperialist structures. We wanted to destabilize, to increase the contradictions.[27]

By June 1972, one month after the attacks in Frankfurt and Heidelberg, a number of the RAF leaders had been caught and imprisoned: Andreas Baader, Holger Meins, Jan Carl Raspe, Ulrike Meinhof, Gudrun Ensslin, Brigitte Mohnhaupt, Irmgard Möller, Klaus Jünschke, and Gerhard Müller.[28] As a result of this sudden defeat, the Revolutionary Cells was founded in 1973 with the intention of pursuing a different type of politics. In one of their early declarations, the RZ announced that they accepted the united front with the RAF; however, because of repression and the "US military strategy," they did not orient themselves toward the RAF.[29] It was decided that the best protection from state repression was to operate in complete anonymity; there were no "RZ representatives" as there were representatives of the RAF.[30] They were considered "*Freizeit-Terroristen*" or "*Feierabendterroristen*," meaning that they pursued terrorism not as their primary vocation but rather during their free time, after returning home from a normal day at work, as it were; hence, they were originally deemed less important than the RAF or less dangerous. This allowed their members to take part in discussions on the left and other legal activities and also prevented a "*Verselbstständigung militärischer Politik*" – an automatic turn to militarized politics.

The RZ did not intend to build an avant-garde organization, nor did they want to represent others. "*Alle müssen alles können*" ("Everyone should be able to do everything") was their goal. There were several other differences between the RAF and RZ. Whereas the RAF wanted to mobilize the German population against the U.S. bases and to attack the United States, just as the United States was attacking targets in Vietnam, the RZ wanted to illustrate that Germany was complicit in U.S. foreign policy. During the early 1970s, the Red Army Faction justified their attacks against the American military presence in Germany because it was an important hub for the military intervention in Vietnam, and not because the U.S. presence was more fundamentally intended to contain, and if necessary, attack the Soviet Union.

Even after the Vietnam War had ended, however, U.S. bases in Germany continued to be targeted by armed groups. In December 1976, the Revolutionary Cells carried out an attack against the Officers' Casino in Frankfurt. In their letter claiming responsibility for the attack, it became clear that the RZ justified their actions in a slightly different way than the Red Army Faction:

[27] Translation by the author; Tolmein 1999: 63–64.
[28] Früchte des Zorns 1993: 79.
[29] Declaration cited in Rabert 1995: 1999.
[30] Nevertheless, the RZ did have "leaders." These included Hans-Joachim Klein, Johannes Weinrich, Wilfried Böse, Brigitte Kuhlmann, and Gert Albartus. See Pflieger 2004.

American officers and generals should no longer feel safe in their casinos in Tel Saatar and Entebbe.... The reason why the BRD is so stuffed full with US military, US capital, and secret services is because they feel so at home here. US imperialism feels safer here than anywhere else in the world, here in its main branch, the imperialist BRD.... This means: imperialist culture is the culture of death. Anti-imperialism is our cultural revolution.[31]

A month later in January 1977, the Revolutionary Cells carried out another attack, this time against the American base in Giessen, located approximately half an hour from Frankfurt:

Today the revolutionary cells blew up the gasoline tank on the military facility in Giessen. Thus the smooth functioning of the US military machine in Giessen was interrupted.... The FRG is not only the main military and economic base of the United States, it is also a moral hinterland, in which the US can recover from the defeats, the revolutionary opposition, the hate which it has encountered in the rest of world. Here US imperialism can regenerate itself.[32]

It is obvious here that the RZ is referring to Vietnam, and that, in their understanding, Germany provided a safe haven for the U.S. "military machine" and was therefore complicit. They were also turning the official justification of the American military presence on its head: instead of the United States offering protection to West Germany, the RZ was insinuating that the United States was seeking refuge and protection in Germany so that it could "recover" from the defeats and setbacks it had experienced elsewhere in the world. The Revolutionary Cells seemed to have set themselves the goal of ensuring that the United States could not, in fact, recover. In June 1978, the American Officers' Club in Wiesbaden was attacked:

After the Officer's casino in Frankfurt was destroyed by our attack last year, many officers have moved to more "secure" casinos, as for example in Wiesbaden. We followed them![33]

A few months later, the new Army barracks that were being built in Garlstedt were attacked. As a result of the Nunn Amendment from 1974, the number of U.S. support forces in Europe was to be cut and the number of combat troops was to be increased within two years. As a result, two new combat brigades were sent to Germany. One brigade was sent to Garlstedt in northern Germany near the supply harbor of Bremerhaven, and the second combat brigade was stationed at Wiesbaden Air Base.[34] Among the army's many construction projects, the one in Garlstedt was of particular significance because it belonged to the British sector. The new base represented an expansion of the U.S. presence into a geographical area over which it had no administrative authority. (Other forms of protest against the base in Garstedt will be discussed later.) In terms

[31] Früchte des Zorns 1993: 370.
[32] Ibid: 371.
[33] Ibid: 372.
[34] Duke 1989: 68.

of the RAF, the attack was significant because for the first time it was justified because of the undemocratic nature of the U.S. presence. The declaration said that the barracks were attacked because they were being built despite the fact that more than 45,000 people signed a petition against it being built, and that 1,500 qm of heathland were being destroyed. This episode illustrates the contradictory position of the "protectariat": they are citizens of their nation-state but subjects of a foreign protection regime.

Despite the death of leading RAF members including Ulrike Meinhof, Gudrun Ennslin, Andreas Baader, and Jan Carl Raspe, the RAF continued to grow and plan ever more astonishing attacks. Whereas the RZ had been carrying out attacks against U.S. interests throughout the 1970s, for several years the RAF had turned its attention to German officials and institutions. The RAF decision to turn its focus back to the American military presence was perhaps at least in part a result of the setbacks of 1977, when the West German government refused to bend despite the kidnapping of leading businessman Hanns-Martin Schleyer; their demand to release imprisoned RAF members was rejected. The 1979 bomb attack in Belgium against then-NATO commander Alexander M. Haig Jr. was the first graphic example of the new strategy. It missed his car by seconds.[35]

The early 1980s witnessed another wave of attacks against U.S. military facilities. Table 3.1 lists violent attacks against U.S. interests in West Germany during 1981.[36] As Table 3.1 illustrates, for the most part, the attacks were not intended to cause harm to individuals, but to damage property. The attempt to assassinate General Kroesen on September 15 is of course a noteworthy exception. After the attack on General Haig in 1979, the assault on General Kroesen represented a second dramatic attempt to assassinate a high-ranking U.S. military officer, as General Frederick J. Kroesen was the commander in chief of the U.S. Army Europe and commander of the Central Army Group (CENTAG) in NATO. The RAF therefore considered him as controlling "imperialist politics from Western Europe to the Persian Gulf."[37] Although the rocket-propelled antitank missiles pierced the car's trunk, General Kroesen and his wife were unharmed. They had been driving an armored vehicle.[38]

The bomb explosion in Ramstein two weeks earlier was also significant because it included a general call to wage war against the U.S. military presence. The RAF letter that claimed responsibility for the event included the call to arms: "Attack the headquarters, the bases, and the strategists of the US military-machine!" However, by referring to Ramstein as a "launch pad for

[35] CSM June 3, 1982.
[36] Information taken from the "Historical Review January 1–December 31, 1981," Headquarters U.S. Army, Europe and Seventh Army, prepared by the USAREUR Military History Office.
[37] Cited in Pflieger 2004: 129.
[38] Nelson 1987a: 168.

TABLE 3.1. *Attacks against U.S. Military and Civilian Facilities in 1981*

Date	Location	Injuries
February 2	Bombs discovered in two OH-58 helicopters in Büdingen	
February 21	Explosion at Radio Free Europe/Radio Liberty in Munich	8 injured
February 25	Molotov cocktails thrown over fence at Ludendorf Kaserne in Kornwestheim, near Stuttgart	
March–April	Firebombing at Amerika Haus Library in Hannover	
	Vandalism at American Express Travel Agency in Heidelberg	
	Firebombing of the American International School in Duesseldorf	
	Firebombing of the military police office in Frankfurt	
	Attempted bombing of U.S. Memorial Library in Berlin	
March 29	Bomb explosion at military intelligence building in Giessen	
March 30	Firebombing of Civilian Personnel Office in Frankfurt	
May 15	Attempted bombing of U.S. Army-leased Bahnhof Hotel at Osterholz-Scharmbeck	
May 19	Attempted bombing of Wiesbaden community headquarters	
May 24	Firebombing of U.S. Army mess hall under construction in Frankfurt	
August 19	Bombs outside U.S. and British installations in Berlin	
August 31	Major bomb explosion at Ramstein	15 injured
September 1	Four cars and a motorcycle were set afire in a U.S. district of Wiesbaden	
September 13	Firebombing of U.S. Consul's residence in Frankfurt	
September 15	Attempted assassination of General Kroesen	
September 16	Attempted bombing of a rail spur at Rhein-Main Air Base	
September 25	Three cars owned by U.S. soldiers were destroyed by arson in the Robinson Barracks in Stuttgart	
October 22	Attempted attack on the home of the V Corps Commander in Bad Vilbel	
November 16	Seventeen cars owned by U.S. military personnel had their tires slashed in Frankfurt	
December 7	Explosive device was tossed through a window of the commander's office of the 5th U.S. Army Artillery Detachment at Bueren	

cruise missiles" and the "headquarters of atomic war in Europe," the RAF was also hoping to appeal to the more militant spectrum of the peace movement.[39] In what army historians called the "most spectacular terrorist act of the year," U.S. Army Brigadier General James L. Dozier was kidnapped from his home in Verona, Italy on December 17, 1981. The kidnapping was organized by the RAF's sister organization in Italy, the Red Brigades (Brigate Rosse).[40]

In 1981, a debate ensued in the West German press about plans to send a special antiterrorist task force to West Germany, which until then had only operated in the Panama Canal Zone. The special forces had been trained in tactics for combating urban guerrillas and were said to have studied the strategies of the Red Army Faction and Revolutionary Cells.[41] In addition, an Anti-Terrorism Operations Center (ATOC) was formed and the USAREUR intelligence liaison officer had a direct line to the chief of the anti-terror agency in West Germany.[42]

According to U.S. Army records, terrorism was at its highest level in 1982, with a total of sixty-eight attacks against the U.S. military presence.[43] Whereas during the previous year terrorists had attempted to assassinate or kidnap four-star U.S. military generals, no such attempts were undertaken during 1982. Toward the end of the year, a wave of car bombings led U.S. officials to fear that they were dealing with a new type of terrorist threat which was no longer targeting military property or high-ranking officials, but aimed to kill indiscriminately. The official Army Historical Review called these "trench-warfare terrorist acts" which were designed to "wear down the command's defenses and morale."[44] U.S. Army personnel assumed that the heightened security measures around major installations and the fact that bodyguards were now carrying submachine guns, may partially explain the turn to "soft targets" or those personnel and facilities which were not as heavily guarded. Social scientists who have analyzed the waves of heavy violence in Germany have also detected a shift from property damage during the early phase to violence against people during the later phase. From 1968 to 1973, only 8 percent of all terrorist incidents involved violence against people; whereas from 1974 to 1977, 50 percent involved violence against people. During the second wave of violence in the 1980s, a similar pattern was found: from 1980 to 1984 only 4 percent involved violence against people, but from 1985 onward, 25 percent involved violence against people.[45] Others have argued that the shift to indiscriminate violence was because of the different leadership of the RAF during

[39] Langguth 1983: 210.
[40] "Historical Review January 1–December 31, 1981," Headquarters U.S. Army, Europe and Seventh Army, prepared by the USAREUR Military History Office: 101.
[41] Nelson 1987a: 171.
[42] Historical Review 1985.
[43] Historical Review 1982–1983: 118.
[44] Ibid: 111.
[45] Koopmans 1993: 644.

the mid-1980s, known as the third generation of RAF cadres. In 1985, Edward Pimental, a U.S. soldier who belonged to the lower rung of the rank-and-file, was murdered by RAF member Birgit Hogefeld to obtain his uniform and I.D. and gain access to the Rhein-Main Air Base. In a public letter, the RAF justified their decision to murder him because he had volunteered to join the army to earn more money:

For us US soldiers in the BRD are not both victims and perpetrators; we don't have this social worker perspective … everyone must realize that this is about war, and they must decide which side they are on.[46]

Whether the increase in indiscriminate violence was because of heightened security measures, different RAF leadership, or some internal logic of escalation cannot be determined by this brief summary. What is perhaps a more interesting question is how U.S. officials came to terms with the fact that Germany, which had become the favorite overseas post for generations of American G.I.s, was now a breeding ground for terrorist networks that posed a challenge to their authority and ability to operate. The three main centers of terrorist activity – Berlin, Frankfurt, and Heidelberg – were, incidentally, also home to some of the most important U.S. military headquarters in Europe. Although the attacks on Generals Haig and Kroesen, two of the highest-ranking commanders on the continent, were certainly alarming, the turn to soft targets posed a problem as well. After all, USAREUR was responsible for more than 1,500 civilian and military installations and guarding all of them was no small task. Furthermore, the 250,000 U.S. personnel were accompanied by an equal number of dependents, meaning that half a million Americans, many of whom lived off base, had to be protected. Some feared that it may be necessary to send all family members back to their homes in the United States, which would have led to a "fortress status" of the U.S. presence.[47] Although this never happened, a certain "bunker mentality" developed among many Americans. Furthermore, as much as possible, U.S. military personnel were to assume a low profile. For example, General Kroesen moved from his private home in Stuttgart to a new residence at the Campbell Barracks. The green military license plates of all American-owned vehicles were replaced by white license plates that resembled those used by civilian Germans, making it more difficult to identify vehicles owned by Americans. For some, all of this was not necessarily a bad thing as some commanders allegedly used the terrorist threat to challenge the troops to stay the course and demonstrate their soldierly virtues. For others, however, realizing that one lived in a potentially hostile environment was surely an alienating experience. After the devaluation of the dollar led many to return to the barracks, purchasing goods at the PX and spending their free time at recreational facilities on base, the ongoing

---

[46] Hoffmann 1997: 344.
[47] Historical Review 1985.

violent attacks meant that the number of interactions between Americans and Germans decreased even further.

By creating an environment of fear, the RAF and RZ can be said to have substantially restricted the off-base activities of some U.S. personnel. However, they cannot really claim any victory for their larger goals. By the mid-1980s, and in particular after the murder of Pimental, the RAF had lost nearly all of their remaining outside support. In an open letter published in the left-leaning *tageszeitung* (die taz), the murder of Pimental was described as a betrayal of everything that had originally motivated the militant left in West Germany. "Your violence has become part of the problem, not part of the solution."[48]

Although the RAF and RZ are often lumped together as left-wing militants, there were significant differences between the two groups. In the interest of better understanding these groups, it is necessary to probe what may appear to be nuances. Whereas the RAF was accused of behaving like soldiers themselves and establishing an elite cadre that paraded as revolutionaries, the members of the RZ operated as average citizens and avoided the spotlight. Furthermore, the RAF was attacking the U.S. military bases and kidnapping and even murdering G.I.s, although the RAF gained important information, which they used during the attacks, from G.I.s who had deserted the military. For example, it was from American G.I.s that they learned about the location of weapons depots and also learned that the IG-Farben house in Frankfurt was not just the headquarters of the V Corps, but also the headquarters of the CIA. Whereas the actions of the RAF were intended as anti-imperialist acts of defiance and hence tinged with a nationalist flavor, the actions of the RZ were meant to highlight the complicity of the Federal Republic. During the 1970s, both groups justified their attacks by pointing to U.S. foreign interventions in Third World countries; it was not until the early 1980s that these groups began to justify their attacks because the U.S. presence was anti-democratic or causing environmental damage, as in the case of Garlstedt. Curiously, despite their murderous tactics, neither the RZ nor the RAF explicitly demanded the withdrawal of U.S. troops or really questioned the fundamental fact of the U.S. presence in Germany. When even those who consider themselves as belonging to the most radical fringes of society do not question this "social fact," it is clear that the U.S. presence in Germany had been sanctified to the point of being inviolable. This began to change in the early 1980s.

The Double Track Decision and the U.S. Presence

On December 12, 1979, NATO approved the plan to deploy a new generation of Pershing II and ground-launched cruise missiles in Western Europe in late 1983, if, in the meantime, no agreement could be reached with the Soviet Union to remove its missile forces from Eastern Europe. The 464 cruise missiles

---

[48] "Offenen Brief" in *Tageszeitung* on September 13, 1985, cited in Pflieger 2004: 149.

were to be deployed at five NATO partners, and all of the 108 Pershing II missiles were slated for deployment in Germany. This decision is usually cited as having triggered the largest social movement in postwar German history. It was also the first time since the end of World War II that the left was able to mobilize more people than the right.[49] In the German literature, this social unrest is generally referred to under the heading of the *Friedensbewegung*, whereas the English-language literature refers to these protest activities interchangeably as a peace movement, an anti-nuclear movement, or even more specifically, as an anti-missile movement.[50] In my assessment, however, these categories are all somewhat misleading because the social movements in Germany during the early 1980s were not directed against any specific ongoing or imminent war – as during the Vietnam Era – but instead against the war-making capabilities of NATO and the United States in particular, that existed on German territory. In 1983 alone, U.S. and NATO troops were the target of more than 400 protest events, including vigils, human chains, blockades, and other types of demonstrations.[51] I argue that this unrest is best understood neither as an anti-war movement nor as a single-issue movement against a particular type of weapon, but that it involved a more fundamental critique of and challenge to the American military presence in Germany than even the allegedly more radical or militant RAF and RZ.

Despite the importance of the 1980s, there are very few English-language studies of the U.S. presence during this period compared to the large literature on the early postwar era. A few exceptions are Daniel J. Nelson, John P. Hawkins, and Anni P. Baker. Hawkins and Baker focus exclusively on problems that were internal to the U.S. military rather than issues relating to the German host country. Curiously, their assessment of the 1980s is quite different. Hawkins emphasizes the strains of living on the front-line of the East-West conflict and the sense of alienation many military families felt, Baker argues that the 1980s were a "golden time" for the U.S. military in Germany because of pay increases, improved housing, and the higher quality of recruits who had freely chosen to enlist in the army.[52] In Nelson's *A History of US Military Forces in Germany*, the only history of the U.S. presence that covers most of the Cold-War period, the NATO double-track decision, is primarily discussed as an elite issue; and in his second book, *Defenders or Intruders*, nuclear issues are not discussed at all.[53] In short, the English literature on this decade is scarce and

---

[49] In the 1960s, the *Vertriebenenverbände* (Associations of displaced people) had meetings with 400,000 people. But by the late 1970s and early 1980s, the left could mobilize twice that many. However, this could in part be attributed to the fact that the issues at stake went beyond those of the traditional left. Fülberth 1999: 204.

[50] I will use all of these terms interchangeably for the sake of avoiding repetition.

[51] This number is taken from the official report of the Office for the Protection of the Constitution (*Verfassungsschutzbericht*) for the year 1983. Cited in Müller 1986.

[52] Baker 2004; Hawkins 2001.

[53] Nelson 1987; Nelson 1987a.

insular in perspective, lacking any consensus on even basic issues. In contrast to this, the German-language literature on the U.S. presence from the 1980s focuses on contentious issues such as the environmental impact of the U.S. presence, land acquisition, maneuver damage, low-altitude flights, and issues of sovereignty. Most importantly, however, from the German perspective, the first half of the decade was defined by the wide-spread peace protests.[54] These movements spread to a much larger portion of the German public than the youth radicalism of the 1960s, which was generally limited to the student-age population in major urban centers. The anti-nuclear and environmental movements of the 1980s branched out into both urban and rural areas, including much of the middle class, and eventually found support from major church organizations, professional associations, and the trade unions. In addition to opposition to the missile deployment, the critique of the U.S. presence in Germany included demands by citizens' initiatives (*Bürgerinitiativen*) to reduce the level of noise pollution caused by low-flying military jets, to decreasing the damage caused by maneuvers such as the REFORGER exercises, to demands that true peace could only be attained when all foreign troops withdraw and Germany regains its full sovereignty. The following section focuses on different types of protest against the U.S. presence during the 1980s: street blockades against the deployed missiles, maneuver obstructions during the annual NATO military training exercises, labor unrest of civilian employees, and activities of the Green Party.

Opposition in the Street

During the late 1970s, U.S. defense planners began casting about for ways to restore U.S. hegemony after the battering it had taken during the early part of the decade. The U.S. had not only lost the war in Vietnam, but had also lost some of its influence with its European allies. Under the leadership of Willy Brandt and Egon Bahr, the Federal Republic had initiated a phase of détente known as Ostpolitik. From the American perspective, one of its most stalwart allies was not only asserting its independence but, by seeking reconciliation with Eastern Europe, was also undermining the Cold War status quo based on the confrontation of NATO and the Warsaw Pact. By engaging in another massive arms race and confrontation with the Soviet Union, the United States could begin once again to reassert its hegemony over the Third World and its European allies. General Bernard W. Rogers, the Supreme Allied Commander of Europe during the first half of the 1980s, defended the NATO double-track decision in an article in Foreign Affairs:

Our objective is a stable world environment with reduced levels of arms and forces, yet we must increase our levels today in order to be successful in negotiating their reduction in the future.[55]

---

[54] Lafontaine 1983; Mechtersheimer 1984; Achilles 1987; Spoo 1989; Angerer 1990.
[55] Rogers 1982.

The logic behind increasing the number of nuclear weapons so as to decrease them in the future may have made sense to military commanders, but the public remained skeptical. E. P. Thompson developed an analysis of the bipolar confrontation, in which he argued that both the USSR and the United States were engaging in a "culture of exterminism." Fred Halliday formulated a critique of this analysis, arguing that focusing on the similarities rather than the differences between the superpowers prevented a complete understanding of the nature of the conflict. In his book *The Making of the Second Cold War*, Halliday analyzed the period from the late 1970s to the mid-1980s. According to Halliday, there were a number of similarities between the first and second phases of the Cold War: the attempt to solidify the military alliances, a massive arms build-up with an emphasis on nuclear weapons, and the suppression of realistic information about the other side. The main difference between the first and second phases of the Cold War, however, was the American loss of strategic superiority. The relative decline of U.S. military superiority was not, according to hawkish defense planners, a result of the defeat in Vietnam, but rather interpreted as resulting from an alleged Soviet military buildup. Hence, phrases like "restoring the balance" were used to portray the arms race under Reagan as defensive and reactive. In German, the word "*Nachrüstung*" was invented which roughly means "compensatory military build up" or a buildup which happens after a Soviet buildup has already taken place to compensate for their advantage.

The so-called double track decision on December 12, 1979, affected not only West Germany, but also the UK, Italy, Belgium, and the Netherlands. All of the 108 Pershing II missiles and 96 of the 464 cruise missiles were to be deployed in West Germany, with the remaining 368 spread out over the other four NATO countries. At the same time, it was agreed that negotiations would be undertaken with the Soviet Union to remove their SS-20 missiles, hence making it a "double" decision. The following month, in January 1980, General Gert Bastian, who was commander of the 12th tank division of the Bundeswehr, sent an eight-page memorandum to Defense Minister Hans Apel complaining that discussion in Germany had been biased by terminology which made it seem that the NATO buildup (*Nachrüstung*) was a follow-up to a Soviet advance (*Vorrüstung*). He argued that considering that the United States was disturbed by the Soviet stationing of medium-range missiles in Cuba in 1962, it was likely that the USSR would see the Pershing II as a provocation. Defense Minister Hans Apel tried to silence the general by transferring him to an obscure desk job. Bastian eventually retired from the Bundeswehr and began a second career as a peace activist and member of parliament, later entering the Bundestag as one of the first nationally elected representatives of the Green Party.[56]

[56] Johnstone 1984: 47.

In November 1980, a group of people, including representatives of the older generation of peace activists from the 1950s (including Martin Niemöller and Karl Bechert) as well as representatives of the new peace movement (including Petra Kelly and Gert Bastian) gathered in Krefeld, where they wrote a document that called on the federal government to withdraw its support for the NATO double-track decision. This document, known as the Krefeld Appeal, had been signed by more than 4 million people by 1983. It represented the common denominator or basic consensus of what would become a very large and diverse movement. Andre Markovits has described the highly heterogeneous movement as being composed of five main groupings. Moving from the left to the right end of the political spectrum these included: (1) grass-roots organizations such as the *Bundeskongress Unabhängiger Friedensgruppen* (BUF) comprised of anarchists, anti-imperialists, and anti-militarists; (2) the left wing of SPD, the Gustav Heinemann Initiative included Erhard Eppler and Oskar Lafontaine and others; (3) the Eco-pax alliance *Bundesverband Bürgerinitiativen Umweltschutz* (BBU) chaired by Jo Leinen, which included also Roland Vogt and Petra Kelly, both founding members of the Green Party; (4) Christian organizations, such as *Aktion Sühnezeichen Friedensdienste* (ASF) and *Aktionsgemeinschaft Dienst für den Frieden* (AGDF); and finally (5) Bundeswehr officers such as Gert Bastian and Alfred Mechtersheimer.[57]

It comes as little surprise that this coalition, ranging from Bundeswehr generals to Christian pacifists to autonomous leftists, also displayed an equally wide range of protest tactics. Some of them organized large rallies, others wrote petitions, whereas others engaged in consciousness-raising activities and yet others turned to alternative lifestyles, with novel forms of protest such as hunger strikes or fasting (*Fasten für den Frieden*). Some have argued that it is possible to divide these various groups into two large camps: those who organized large demonstrations to influence politicians, and those who tried to interrupt the functioning of the military itself.[58] The former group is often seen as including the "traditional" spectrum ranging from social democrats to communists to Christian groups, whereas the latter group composes the "alternative" spectrum of independent, anarchist, or anti-militarist groups. Although this classification disregards other types of contentious politics, such as petitions within the parliament, it is useful in that it distinguishes between two divergent conceptions of politics: those who thought that it was possible to bring about change by convincing the authorities through rational arguments and those who thought it was only possible to disrupt their actions. Over time, the independent spectrum of the peace movement became more influential.[59]

Perhaps unsurprisingly, the large peace rallies have attracted the most attention within the social movement literature. The first mass demonstration took

---

[57] Markovits and Gorski 1993.
[58] Lou Marin 2007.
[59] Interview with Otfried Nassauer in Berlin in January 2008.

place on October 10, 1981 with about 300,000 participants in Bonn, and the last large rally of this kind was in Hunsrück with 180,000 people in 1986. The high point of movement activity crescendoed in the nationwide Action Week (*Aktionswoche*) in October 1983. One of the most dramatic protest events during this week was the formation of a sixty-mile-long human chain including some 200,000 people who physically linked together the Headquarters of the U.S. European Command in Stuttgart to the American Army barracks in Neu-Ulm, which was to serve as one of the sites for the Pershing II missiles. During the same week, Erich Honecker, the President of the East German government, sent an open letter to Helmut Kohl, asking him to veto the missile deployment "in the name of the German people."[60]

During this same time, political crises among the elite were developing on two levels: both within the German governing coalition, and within NATO, in particular the bilateral relationship between Washington and Bonn. It is worth emphasizing that one result of the bilateral crisis was the creation of a new political office in 1981, known as the Coordinator of German-American Relations, which still exists today.

In September 1982, the social-liberal coalition broke apart primarily because of budgetary disputes; the FDP decided to form a coalition with the CDU/CSU. Chancellor Helmut Schmidt (SPD), who had supported the NATO double-track decision, and who had come under immense pressure from within his own party, was deposed by a constructive vote of no-confidence on October 1, 1982. Helmut Kohl (CDU) was elected chancellor, while Hans-Dietrich Genscher (FDP) remained Foreign Minister. Now in the opposition, the Social Democrats switched sides and joined the growing ranks of the peace movement.

Symbolic Civil Disobedience: Blockades of Military Bases and Missile Depots

The Pershing II missiles were to be deployed at three missile depots: Mutlangen, Heilbronn, and Neu-Ulm. In fact, the turn to civil disobedience within the peace movement began before the deployment of the missiles. Similar to the German trade unions which carry out *Warnstreiks* or warning strikes to show their determination to go on strike should the collective bargaining process break down, peace activists carried out "*Warnblockaden*" to show that they were resolved to prevent the deployment.

On September 1, 1983, demonstrations began outside the two military bases in Mutlangen and Bitburg, where the Pershing II and cruise missiles were to be deployed. The protest in Mutlangen in particular attracted the attention of *The New York Times*, *The Washington Post*, and *The Wall Street Journal*, although with around 5,000 demonstrators, it was much smaller than the mass rallies in Bonn and elsewhere which attracted several hundred thousand participants. The two main reasons for the media attention were because it

---

[60] *The New York Times*, October 24, 1983.

was timed to coincide with Hitler's invasion of Poland forty-four years earlier and also because more than 100 prominent people attended, such as Günter Grass, Heinrich Böll, Petra Kelly, Oskar Lafontaine, and Erhard Eppler, as well as Daniel Ellsberg and Phillip Berrigan from the United States.[61] These well-known politicians, intellectuals, and activists were divided among "affinity groups" (*Bezugsgruppen*), which worked in shifts and succeeded in blockading all traffic in and out of the base for three days. The U.S. military did not attempt to pass the blockade and the public affairs officer for the 56th Field Artillery Brigade, Major Tony Maravola, believed that their restraint would ensure that nothing "newsworthy" would happen. Maravola did concede, however, that the blockade was an inconvenience because base workers were not able to return home in the evenings but had to sleep on cots within the base and all traffic in and out of the depot was halted. If the protest had seriously hindered the activities at the base, however, Maravola assured journalists that the German police would have been ordered to intervene. One mounted German policeman was cited in *The Wall Street Journal* as saying, "We all want peace and would prefer it without the missiles, so why shouldn't we get along?"[62]

The same restraint was not shown at Bitburg Air Base, however, and *The Washington Post* reported that water cannons and dogs were used to disperse about 800 protesters, and 150 people were taken into custody.[63] The protests in Mutlangen became symbolic of nonviolent civil disobedience, whereas the violence used to remove protesters in Bitburg signaled that the authorities were engaging in a type of "flexible response" toward protesters. In other words, while showing restraint toward the high-profile blockade in Mutlangen, the blockade in Bitburg, which was carried out by rank-and-file peace activists and ordinary citizens, were faced with a higher level of state violence and intervention. Some have interpreted this as a divide and conquer tactic, or as an attempt to incite divisions within the peace movement. According to Rainer Trampert, one of the early members of the Green Party, "Mutlangen showed us that the police are only nice to us if we are particularly harmless."[64] Ironically, although the major U.S. newspapers were sure to point out that the protests were only partially successful because the nonviolent response of the police was anticlimactic, the fact that major U.S. papers would pay attention to a relatively small protest in an unheard-of village could be interpreted as a success in and of itself.

Official U.S. Army documents provide insights into how demonstrations such as these were dealt with. Perhaps most importantly, the U.S. military

---

[61] "W. German Protesters Blockade U.S. Base," *The Washington Post*, September 2, 1983 and "Quiet Protest and U.S. Base in Germany," *The New York Times*, September 2, 1983.

[62] "Lukewarm West German Demonstrations against Missiles Could Hinder Movement," *The Wall Street Journal*, September 6, 1983.

[63] "Water Canons, Dogs rout 800 Protesters at West German Base," *The Washington Post*, September 4, 1983.

[64] *The Washington Post*, September 20, 1983.

operated with the concept of non-direct involvement in demonstration control in both Germany and Turkey. In other words, although the United States would provide equipment and facilities that would support host nation counterdemonstration planning, dealing with protest events directly was the job of the German or Turkish authorities. To do otherwise would have created the appearance that the host nation was forfeiting even more of its sovereign responsibilities to the United States than it already had. It also meant, however, that the costs of counterdemonstration efforts were carried by the NATO allies and not the United States. Secondly, U.S. military personnel were ordered "not to document demonstration activity with audio, video, or photographic equipment."[65] This was most likely intended to avoid creating the impression that U.S. soldiers were personally engaging in surveillance activities, but also to cut down on the number of records of such protest events. Thirdly, the U.S. military was forced to accept the fact that blockades such as those described earlier, would disrupt the functioning of some facilities. However, a total blockade of U.S. forces installations was considered "unacceptable." And finally, the issue of when violence could be used in the repression of protest activities was important. Although the German authorities were responsible for policing demonstrations in Germany, it was apparently senior U.S. commanders who could decide when "deadly force" could be brought to bear:

Deadly force was to be an absolute last resort when all other lesser measures failed. Only those senior US commanders expressly authorized to do so by USEUCOM Regulation (UR) 10–5 could designate property or information (sic) as being vital to national security, this designation being the prerequisite for employment of deadly force.[66]

The fact that deadly force was considered a legitimate means of protecting property and information illustrates the fact that the state intended to monopolize not only violence, but also certain types of information or knowledge. The rise of peace research institutes and counter-experts hence represented an explicit challenge to the state's monopoly on information. This was a particularly sensitive issue in terms of nuclear weapons. In the words of Breyman, "peace movement 'counter-experts' despoiled the church of the nuclear priesthood."[67]

"In Case of Setback, the Government Should Dissolve the People and Elect A New One"[68]

According to opinion polls, nearly 75 percent of the German population was opposed to the missile deployment. Nevertheless, on November 22, 1983, the Bundestag approved the missile deployment by a vote of 286 to 225. Less than

---

[65] Historical Review 1982–1983: 136.
[66] Ibid.
[67] Breyman 1998: 195. Two of the most important centers for peace research include the Berlin Informationzentrum für transatlantische Sicherheit (BITS) and the Starnberger Institut für Friedensforschung.
[68] From Bertolt Brecht's poem "The Solution."

twenty-four hours after the Parliament approved the deployment, the first bat-
tery of Pershing IIs was reported to have been flown to an air base, causing
many to speculate that Washington never seriously doubted that the resolution would be passed.[69] In mid-October, more than a month before the vote in
the Bundestag, Green parliamentarians brought to the attention of their colleagues that equipment belonging to the Pershing II missiles was already stored
in an army facility in Frankfurt-Hausen.[70] Finally, the fact that Richard Perle
and others had made public statements about how Germany was an occupied
country and hence the United States could station whatever it wanted there,
did not do much to placate the already tense situation.[71]

An internal report entitled "Why the Germans are so concerned about
the prospect of nuclear modernization" that was distributed to the General
Staff and written in the summer of 1983, provides insights into how the army
viewed the opposition to the missile deployment. The report was confident that
the Bundestag would approve the deployment, but less confident that the associated societal discontent would then go away.

There is no reason to believe that the peace movement will give up when the first
Pershings arrive. We must assume that from now on, one of the major political parties,
the SPD, will oppose nearly everything we want to do locally, regionally and nationally.
The Greens and the churchmen and the intellectuals are not going to change their minds
either. And the Chaoten and the Communists will continue to work to unravel civilized
society and the NATO alliance. We would be well advised to think of how in the longer
run we are going to live, operate and train in a Germany where the bi-partisan consensus on defense matters has broken down and where our commanders are going to be
spending a great deal of their time figuring out ways to operate without giving offense
to the civil population. That, in the long run will be the major impact on USAREUR.[72]

As we will see, this report proved to be quite prescient. Despite the attempts
to criminalize some of the activists, and despite the fact that the major news
media was declaring the failure of the movement, some decided to become
"Dauer-Blockierer" or permanent blockaders and gave up their careers or university studies and moved to the rural village of Mutlangen, where they set up
operations. These permanent blockaders built themselves a base of their own,
now known as the Carl Kabat House. From here they planned the daily blockades that would take place and also eavesdropped on the police by intercepting
their radio signals, an activity they called "counter-surveillance."[73]

[69] November 24, 1983, *The New York Times*.
[70] The boxes that allegedly contained the equipment were marked with the words "Pershing
Modification Team" and "Pershing Cylinder Assembly." "Kisten, Raketen und die Staatsgewalt"
Die Zeit, October 21, 1983."
[71] Cited in Johnstone 1984: 59.
[72] "Why the Germans Are So Concerned about the Prospect of Nuclear Modernization," internal
report by Benton Moeller, Summer 1983; private collection.
[73] Interview with Wolfgang Schlupp-Hauck in Mutlangen 2006.

At any given time, one-third of the Pershings were in maintenance, one-third were deployed in the forest, and one-third were kept in what was called "quick reaction alarm." As the missiles would be driven out from the base in the morning and returned in the evening, this meant that a permanent rotation of the missiles was taking place. Nuclear convoys driving from one end of the town to the other, passing residential areas, schools, and playgrounds, became a part of everyday life.

A small group of activists formed the hard core of permanent blockaders; however, they organized a series of blockade actions, each with a specific theme, that attracted large numbers of people to Mutlangen. Some of the more noteworthy of these blockades included the: *Adelsblockade* (nobility blockade), the *Erntedank-Blockade* (Thanksgiving blockade), the *KZ-Häftlinge-Blockade* (the concentration camp survivor blockade), the *Konzert-Blockade* (the concert blockade) – which attracted classical musicians from all over Germany who performed a concert in front of the missile depot, the *Senioren-Blockade* (senior citizens blockade), and the *Richter-Blockade* (Blockade of Judges). These blockades ensured that a constant stream of activists was descending on the village on a regular basis, and in some ways the trip to Mutlangen became a type of rite-of-passage.

Although the initial reaction to the Celebrity-Blockade in Mutlangen was much more tolerant than the response of the authorities in Bitburg, a wave of court cases began against the blockaders in February 1984. Over the course of the next few years, 2,999 people were arrested in Mutlangen; 95 percent of the cases of those who were tried for the blockades were accused of "*Nötigung,*" which was interpreted as a type of coercion. In the court, the allegation was that the peace activists had prevented a third party ("unbeteiligte Dritte") from "*persönlichen Fortbewegungsfreiheit.*" In other words, the crime that the blockaders had allegedly committed was that they had hindered the soldiers in their freedom of movement. In this way, the state had redefined nonviolence as violence. The peace activists argued that the soldiers were acting on orders, which had nothing to do with their own personal choice or their freedom of movement. In fact, the peace activists were only attempting to cause the soldiers to pause and reconsider their actions. "Instead of acting as someone who is merely carrying out orders as a *(Befehlsempfänger),* and as someone who is trained to kill on a large scale, the soldier stops in front of a group of unarmed people and behaves like a 'normal human being' which respects the right to life of others."[74] The activists wanted to create "*Tötungshemmung*" – an unwillingness to kill – which was diametrically opposed to the principle of the military. They wanted the soldiers to become unsure of their soldierly duties and to question the use of nuclear weapons. According to the activists, they were in a win-win situation: if during a blockade they were not removed from the street, then they had successfully prevented the movement of the missiles. If,

[74] Nick, Scheub, Then 1993: 164.

however, they were carried away by soldiers, then they would disturb the soldiers and make them question what they were doing, which was also considered a success.

"We Like Your Face, Not Your Base"[75]

In an open letter entitled "To the American Soldiers stationed in Germany," the blockaders clarified their position regarding the individual G.I.s who were both in Mutlangen and other parts of Germany. The letter, written in English, began with a quote from Martin Luther King which argued that the standoff between the superpowers had created such a dangerous situation that "in this world nonviolence is no longer a matter of theoretical investigation but rather a commandment to act." The letter continues by addressing the American soldiers directly:

> We want you to know that our actions are NOT aimed at you personally. Also, our actions are NOT anti-American – our actions are anti-Cruise Missiles and anti-Soviet SS 20! One group of blockaders will be US citizens! We want you to respect our physical integrity when we are sitting in front of your vehicle. We have made up our minds that it is better to go to jail as a consequence of nonviolent action, than to violate our conscience. Disarmament will only be possible if everybody does what he or she can do to achieve it. Today we ask you. How do you work for disarmament?[76]

To what extent the blockaders achieved their goal in triggering some form of self-questioning among G.I.s is unclear. Since the draft had been abolished in 1973, insubordination and resistance within the military had become a relic of the Vietnam era and no longer characterized the professional army. However, occasionally peace activists were able to convince individual G.I.s. to act as witnesses during their court case and attest that they had not been harmed by the blockades.[77] On the whole, however, the G.I.s of the 1980s who had volunteered for military service proved fairly immune to the politicization attempts of the activists, yet within German society, the tactics of civil disobedience became more and more accepted over time.

Between 1984 and 1987, more than 30,000 people participated in blockades and demonstrations in Mutlangen. One of the largest single demonstrations was during the traditional Easter March in 1984, when more than 20,000 people formed a human chain encircling the base.[78] When one considers that Mutlangen was a small village with a population of less than 6,000, it becomes clear that Mutlangen represented not only the turn to civil disobedience, but also the geographic spread of the peace movement from urban to rural areas. According to police records, 2,999 people were arrested under the

---

[75] This was one of the slogans that the activists used to convince the G.I.s that their protest was anti-base, not anti-American.

[76] Nick, Scheub, Then 1993: 166.

[77] Interview with Wolfgang-Schlupp Hauck in Stuttgart on January 18, 2008.

[78] *The New York Times*, April 24, 1984.

Nötigungsparagraph, 200 of whom served time in prison while the others paid fines that were determined by their salary.[79] The protests in Mutlangen culminated in the Blockade of Judges in 1987. This is also an indication that the initial attempt to criminalize those who were protesting at Bitburg did not work; although this tactic may have initially discouraged some from participating in civil disobedience, and although not all groups within the peace movement supported direct action, over time this type of tactic became more acceptable, as even state officials such as judges took part. The Intermediate-Range Nuclear Forces (INF) Treaty in 1987 represented a belated victory for the blockaders in Mutlangen. Finally, after the end of the Cold War, the *Nötigungsparagraph* was deemed unconstitutional in 1995.

Modernization, Wartime Host Nation Support, and the Master Restationing Plan

An under-reported aspect of the missile deployment is that it was but one element of a much larger transformation and modernization of U.S. forces in Germany. Beginning in 1980, the U.S. Army undertook one of the most ambitious modernization programs in its history. In addition to the Pershing II and cruise missiles, a whole slew of other weaponry was modernized or introduced for the first time: the AH 64 helicopter, the multiple launch rocket system, the M1A1 tank, the infantry fighting vehicle, and the Patriot air defense system. In addition to introducing new technologies, the transformation of the U.S. presence during the 1980s also included provisions for emergency planning with far-reaching political implications. On April 15, 1982, a new agreement known as the Wartime Host Nation Support (WHNS) Treaty was signed, which had been preceded by a series of secret negotiations. The agreement allowed the United States, in a crisis situation, to effectively double the size of the army and air force troops present in Germany. It also allowed the German military to play a supporting role in such a crisis situation by training and equipping up to 93,000 Bundeswehr reservists to support the U.S. forces. Under the agreement, ammunition and equipment for the additional troops would be stored in the Federal Republic so that they would already be in place when the reinforcements arrived. Finally, it provided for the stationing of aircraft at more than seventy collocated operating bases (COBs) in Europe, seventeen of which were in the FRG, which under normal circumstances were based in the United States.[80] Critics claimed that if implemented, the WHNS could undermine the democratic principle of subordinating the armed forces to civilian rule because it involved turning over large parts of the civilian infrastructure to the military. For example, the U.S. military estimated that it would need 20 percent of the German train system's capacity (*Bundesbahnkapazität*) if the WHNS were

---

[79] Interview with Wolfgang-Schlupp Hauck in Stuttgart on January 18, 2008.
[80] Duke 1989: 85.

implemented. Provisions were also included in the treaty that would have allegedly allowed U.S. officials to influence radio and television programming.[81]

When confronted with German opposition to American demands, one familiar tactic was to use the threat of U.S. troop withdrawals as a type Damocles
sword. For example, during the early months of 1982, the U.S. Ambassador in
Bonn, Arthur F. Burns, issued two warnings that Washington could pull out its
troops from the Federal Republic. Defense Secretary Weinberger at the Munich
Security Conference in 1982 mentioned the same possibility. Against the background of the grandiose schemes that defense planners had envisioned in the
early 1980s, however, the "threat" of troop withdrawals was becoming ever
harder to believe. Karl Bredthauer, the editor of *Die Blätter*, gave voice to this
skepticism. In his editorial, he wrote that pulling out the troops and nuclear
missiles from Germany would mean that the most urgent demands of the peace
movement would be realized, but that the Americans would never dream of
doing such a thing. Instead, they were planning the "Second Occupation of
Europe" (*Zweite Besatzung Europas*) in the form of the Wartime Host Nation
Support Treaty. This "second occupation" was happening thirty-seven years
after the end of World War II, twenty-seven years after Germany became sovereign, ten years after the ratification of the renunciation of violence with the
East, and nine years after both German states joined the UN.[82]

Last but not least, the transformation process included the plan to radically
alter the overall positioning of U.S. forces. According to defense planners, the
U.S. presence was located too far from the border to East Germany. The grand
design of relocating the largest overseas American military presence became
known as the "Master Restationing Plan" (MRP). According to the Historical
Review from 1981: "From the German side, the reaction to the MRP aspect
of the Stoessel Demarche was uniformly negative. FRG Chancellor Helmut
Schmidt rejected the entire demarche."[83] Schmidt, it should be remembered,
had supported the missile deployment in the face of massive opposition within
his own party; hence, the fact that he did not support the MRP was significant. High-ranking military officers held a similarly dim view of the proposal.
General Jürgen Brandt, Chief of Staff of the Armed Forces, even argued that
there was no military necessity for the move eastward because troops could easily and quickly reach the East German border from their existing positions.[84]
In sum, whereas the missile deployment, force modernization, and Wartime
Host Nation Support Treaty were all controversial as they engendered various
degrees of public protest, the Master Restationing Plan was nipped in the bud
as it was rejected outright by the German chancellor and Chief of Staff. The

[81] Schmidt-Eenboom unpublished manuscript.
[82] This is one example where the rhetoric used by German activists was similar to the Turkish rhetoric that framed their opposition to the U.S. presence as a second national liberation struggle.
"Ziehen die Amerikaner ab?," Bredthauer in *Die Blätter*, March 1982, JG 27: 261.
[83] Historical Review 1981: 48.
[84] Ibid.

opposition to the U.S. presence, which had begun in small activist groups on the margins of the political scene, had now reached the highest echelons of West German politics.

Throughout the Sturm und Drang of these controversies, the U.S. Army and Air Force continued to carry out normal training exercises as well as large-scale military maneuvers. As discussed previously, the Federal Republic of Germany was one of the most heavily militarized territories in the world. The German *Weissbuch* from 1983, published by the German Ministry of Defense, underlined the FRG's contribution to the NATO alliance: "In no other western country does so much military training take place on such densely populated territory."[85] The reason for the importance of German territory as a training ground was clear. In the event of war, NATO planners assumed that there were three potential invasion routes for Warsaw Pact forces, all of which were located in Germany: either through the Northern German plain, or the Bavarian *Mittelgebirge*, or through the so-called Fulda Gap. Regardless of which route was chosen, Germany would have inevitably become the main battlefield between East and West. Therefore, as the first-line of defense within Europe, Germany had been declared by NATO to be a "combat zone." It was divided into the "Forward Combat Zone" (FCZ) near the border to East Germany, further west was the "Corp Rear Area," and then the "Rear Combat Zone" (RCZ) in the westernmost part of southern Germany. Major facilities such as the Ramstein Air Base as well as storage facilities and weapons depots were located there.

This meant that during peacetime, the territory and the entirety of the civilian infrastructure of the Federal Republic could potentially serve as an arena for military training purposes. Hence, although the last real battle was fought on German soil in the spring of 1945, during the nearly four decades of the Cold War, NATO troops were kept in a permanent state of readiness and military maneuvers were carried out on a more or less ongoing basis: training was pursued on a twenty-four-hour-a-day basis 363 days a year. Only Christmas and Easter were exempt from firing. According to the U.S. Army, this schedule was necessary because of the large number of troops that needed to be trained.[86] Those communities near the major training areas of Grafenwöhr, Wildflecken, and Hohenfels were especially affected by the constant firing and training exercises. Although German citizens had been complaining about this type of noise for many years, by 1982 their complaints had secured the support of several high-ranking political officials including Minister-President Strauss (Bavaria), Börner (Hesse), Federal Finance Minister Stoltenberg, and Parliamentary State Secretary Wurzbach. Minister Stoltenberg requested that the U.S. German Environmental Committee explore alternatives and make recommendations.

[85] German Federal Ministry of Defense, *Weissbuch*, 1983.
[86] Historical Review, HQ U.S. Army Europe and Seventh Army: 280.

Although the group met three times in 1983, they failed to reach a consensus on a reduction of firing schedules and advance notification.

During the 1980s, conflicts regarding training exercises arose with the Bundeswehr as well. As noted in the Historical Review of the U.S. Army Europe and Seventh Army, Brigadier General John W. Foss, Commander of the Seventh Army Training Command, detected "the emergence of a very negative attitude from the German side." Apparently the German officers felt that all training areas in Germany should be administered by the Bundeswehr and not the U.S. Army, with NATO headquarters holding scheduling authority over them. General Foss noted that the Germans apparently felt that the Americans were running the training areas as "an occupying power" because many rules were not enforced and because the firing hours did not take the local population into consideration.[87] It should be noted that the rhetoric of an "occupying power" is being used not by marginal groups of radicals, but by German military officers. Furthermore, the Bundeswehr had to spend money to send their own units out of the country to train because of a shortage of training time at in-country training areas. The Historical Review concluded with these words of warning:

It also appeared that the Germans had a long-term goal of obtaining administrative control of all major training areas and would wait until the most advantageous time to present their views.[88]

War Games

In addition to routine training exercises, the U.S. military carried out major maneuvers as well. Every year, the United States participated in around 5,000 maneuvers that lasted three to four days and in which up to 2,000 soldiers were involved. Some eighty maneuvers were even larger, lasting longer than four days and in which more than 2,000 soldiers were involved. The German Bundeswehr participated in approximately half of these maneuvers.[89] Despite the fact that the US alone possessed a formidable amount of real estate in West Germany, the Army did not restrict itself to these already quite extensive properties, but often carried out training exercises off-base, on private property including open fields and farmlands.

In fact, according to Field Manual 100–5 from 1982, the U.S. Army had to be prepared to fight even in the urban areas of Germany. However, the urbanization of Germany posed a major challenge for military operations. Urban areas "restrict visibility, fields of fire, and maneuver capability."[90] In some ways it would appear that the major obstacle for the U.S. Army was that Germany

---

[87] Historical Review January 1, 1982–December 31, 1983, HQ U.S. Army Europe and Seventh Army: 376.

[88] Ibid: 376.

[89] Weissbuch 1983.

[90] Field Manual FM 100–5, Headquarters, Department of the Army, Washington August 20, 1982: 17–26.

was an inhabited country and therefore posed some limitations on their ability to operate.

Of course, extensive training exercises did not begin in the 1980s, but they have a long history. The first NATO exercise which was premised on the use of nuclear weapons was called "Carte Blanche" and was held in June 1955. Although these NATO exercises are merely practice maneuvers used to maintain a high-level of readiness, they are premised on the notion that Germany could become a real battleground for a real war, and – in this particular case – one in which nuclear weapons were used. For this reason, the number of deaths were tallied, just as happens in the aftermath of an actual war. In this particular exercise, the death toll was calculated at 1.5 million dead and 3.5 million wounded. According to a public opinion poll taken after the exercises, only 13 percent of the German population wanted nuclear weapons on German soil. This is just one indication that even during the early period of the Cold War, some Germans did not feel protected by the U.S. protection regime. Public support for these types of war games remained very low, but it was not until the 1980s that these misgivings turned into widespread opposition. The turn to opposition was attributed to a number of reasons, including the more aggressive posturing of NATO after the détente of the 1970s, as briefly described previously, and the fact that the war "games" were causing unprecedented damage and even civilian deaths. A memorandum written for the Commander in Chief, U.S. Army Europe and Seventh Army describes the public mood in the early 1980s:

While environmental awareness in Germany has traditionally been strong, the exponential development of environmental consciousness in Germany, accompanied by a tightening of government regulations on noise, air, and water pollution, has placed the US Forces in a progressively constricted situation regarding both the use of facilities under US Forces control and the use of open land under maneuver rights for training purposes. Another subjective factor which shows no sign of abating is growing impatience by the public with the inconveniences caused by military training activities and exercises. Leftist groups have been partially successful in exploiting this mood by advocating everything from opposition to maneuvers to unilateral disarmament. Nevertheless, it would be an error to write off the current mood as solely leftist, either in origin or direction. I suggest it is more the result of a long period of peace and prosperity in Europe.[91]

The largest annual exercises were the Return of Forces to Germany (known as REFORGER) which were designed to practice the rapid-reinforcement of U.S. troops by airlifting thousands of G.I.s to their pre-positioned equipment. During these exercises, an additional 35,000 U.S. soldiers would join the army

---

[91] "Special Activities Report on Impact of German Industrial, Residential and Recreational Development on U.S. Forces in the Federal Republic of Germany," Memorandum for the Commander in Chief, U.S. Army Europe and Seventh Army from the U.S. Forces Liaison Office, Bavaria, March 31, 1981.

divisions already stationed in Germany, where they would operate on average 50,000 trucks and 10,000 tanks in addition to other armored vehicles, often roaming over thousands of acres of German territory. In this way, large portions of central and southern Germany would become a vast playpen for thousands of NATO troops.

The 1981 edition of the Army Field Manual announced the adoption of a major new military doctrine called AirLand Battle. This doctrine involved coordinating ground and air systems to carry out "deep strikes" or attacks deep inside enemy territory. Dissenting voices within the Bundeswehr have referred to the AirLand Battle doctrine as the "American version of the Blitzkrieg."[92] The peace movement, which was already able to mobilize people in unprecedented numbers since the Double Track decision in 1979, was taken aback by this new doctrine because it was clearly an offensive strategy, although NATO was officially a military alliance aimed at the defense of Western Europe against the Warsaw Pact. The connection between AirLand Battle and the missile deployment planned in 1983 became evident: whereas the Pershing IIs and cruise missiles kept Moscow in check as they were targeting cities and command and control centers in the Soviet Union, NATO could potentially carry out attacks in the Persian Gulf or others areas outside NATO. These fears were then confirmed during the 1986 attack on Libya.[93] This was the first time it became clear that NATO was preparing to carry out attacks outside its sphere of influence, hence potentially upsetting the status quo which had existed for almost forty years.

This new doctrine was put into practice for the first time during the Maneuver called "Scharfe Klinge" in 1981. During the Fall Maneuvers in 1983, U.S. soldiers practiced the digging of mass graves for the first time.[94] In other words, almost thirty years after the Carte Blanche exercises, the U.S. military was still carrying out war games premised on the idea that they would result in a massive amount of civilian casualties. For many, the notion that the "defense" of Germany would necessitate pandemic suffering and casualties became an increasingly macabre calculation. Furthermore, regardless of which strategy was behind the training exercises and regardless of the number of imaginary casualties that the war games were taking into account, these maneuvers resulted in very real widespread damage to both rural and urban areas. One scenario was described in an article entitled "War Games Wreak Havoc in West Germany" in *The Washington Post* from September 24, 1984:

---

[92] Interview with Lieutenant Colonel Jürgen Rose in Munich on January 15, 2008.

[93] The bombing of the Libyan capital of Tripoli was in retaliation for the attack against the La Belle disco in West Berlin on April 5, 1986 in which two U.S. military personnel and a Turkish woman were killed. Sixteen years earlier, Colonel Muammar Qadhafi had evicted the United States from the Wheelus Air Force Base after he came to power.

[94] *Der Spiegel*, October 3, 1983.

Out of the blowing dirt, the din of growling truck engines and the rumble of tanks rip-ping up the narrow streets, a scene from World War II was played out here last week in this tiny Bavarian town. Townspeople looked down silently from their bedrooms to the streets below where six camouflaged West German Leopard 2 tanks manned by soldiers with blackened faces were stalled in a midnight traffic jam. The tanks were blocked by a line of heavy Army supply trucks. The lead truck finally lumbered aside, scraping against the wall of an ancient church, and the tanks roared out of town, masticating the edges of a sidewalk as they went.[95]

Six soldiers died during the maneuvers in 1984 and more than $10 million in damage was done to private property. This type of destruction was so com-mon that each military unit would travel with a claims officer whose job it was to distribute damage-claim forms and cash. To reduce the number of claims that were filed, officers were authorized to settle for up to $70 on the spot. Heavy tanks would rip up fields, creating deep trenches and compact-ing the earth, turning fertile farmland into barren terrain. The cartridges and metal shell casings that were left behind posed a danger for livestock, which could be injured or even die if the cartridges were swallowed and the sharp edges cut into their stomachs.[96] The fact that claims officers accompanied army units indicates that U.S. defense planners were fully aware of the dam-age caused by the war games. In other words, the grievances of the population cannot be dismissed as resulting merely from their own subjective perceptions or political ideologies, but were based on the very real collateral harm caused by the U.S. presence.

As part of the all-encompassing modernization program of the U.S. forces, a new generation of tanks was introduced. Each tank weighed ten tons heavier than the previous generation of tanks. Soil compaction, known in German as *Bodenverdichtung*, became a serious issue for farmers. In addition to making it difficult for crops to grow, heavy tanks and other vehicles could disrupt the drainage and irrigation systems by creating trenches and flattening small hills. Other environmental concerns included the contamination of the water supply and gasoline and oil spills. Not only were farmers and local citizens increasingly exasperated by the maneuvers, so was anyone who made use of the German Autobahn system. During some of the large maneuvers, it was possible that long stretches of the 500 km-long A3 Autobahn, which connected Wiesbaden in southern Germany to Hamburg in the north, could be blockaded by military convoys for hours at a time. Accidents were a common occurrence, not only on the training ranges, but also on the Autobahn, and on public and private property. The table below summarizes the number of accidents, inju-ries, and fatalities during the Return of Forces to Germany exercises in 1982 and 1983.

---

[95] "War Games Wreak Havoc in West Germany," *The Washington Post*, September 24, 1984.
[96] Ibid.

	REFORGER 1982	REFORGER 1983
Accidents	263	191
Military Injuries	138	63
Civilian Injuries	17	18
Military Fatalities	5	2
Civilian Fatalities	7	10
Total Fatalities	12	12

From Symbolic to Disruptive Civil Disobedience: Maneuver Obstructions or "Don't Defend Us to Death!"

Beginning in the early 1980s, some groups who belonged to the independent spectrum of the peace movement were not mollified by the compensation for damages or the notion that this was a necessary price to pay for protection from a Soviet threat that no longer seemed imminent, and began to organize so-called *Manöverbehinderungen*, or maneuver obstructions. Although civil disobedience had already become an established form of contentious politics during the blockades of the missile depots after the deployment of Pershing II and cruise missiles in 1983, maneuver obstructions were considered highly dangerous and were only partially supported by the peace movement and the Green Party. A number of prominent peace activists who had participated in the street blockades in Mutlangen – including Heinrich Albertz, Gert Bastian, Heinrich Böll, Erhard Eppler, Oskar Lafontaine and Helmut Priess – had spoken out against the maneuver obstructions. Annemarie Borgman, who was at that time the Speaker of the Green Party in the Bundestag, criticized the six prominent peace activists who were against the maneuver obstructions, arguing that their critique may further divide the peace movement into *Demonstrierer* (demonstrators) and *Blockierer* (blockaders), which was one of the goals of the state. Borgmann echoed Martin Luther King in her appeal to civil disobedience: "*wo Recht zu Unrecht wird, wird Widerstand zur Pflicht*" (when right becomes wrong, resistance becomes a duty).[97]

Despite the lack of support from segments of the peace and environmental movements, activists gathered in the Fulda Gap region in September 1984, intent on demonstrating that the missile deployment did not mean the end of the peace movement. During these maneuvers, known as the Autumn Forge exercises, peace activists operated from five separate camps to disrupt the maneuvers as well as inform the local population about the harmful effects of the training exercises. Although many were doubtless aware of the most visible forms of damage, some of the less-visible but equally deleterious side effects of the maneuvers were not common knowledge.[98] Because NATO troops relied

[97] "Grüne billigen Manöverbehinderungen," *Frankfurter Rundschau*, August 13, 1984.
[98] "Friedensbewegung: Netz knüpfen," *Der Spiegel*, August 13, 1984.

on the civilian infrastructure, it was possible to create mayhem simply by removing traffic signs or painting over them as to confuse the drivers of military vehicles. Some covered road signs with self-made signs reading "Off-limits to soldiers," parodying the official signs around the perimeter of many military bases. One group recommended destroying antennas and telephone cables so as to prevent communication between officers and troops in the field. Others planned to remove the signs at bridges which indicated the maximum weight of vehicles that the bridge could withstand, so that drivers would be uncertain about whether or not they could cross.

Those who were still undeterred went a step further. As reported in *The New York Times*, for example, 188 people broke into an army training area near the East German border bearing signs reading "Don't defend us to death!" and "Don't abuse our country for your aims." Lieutenant General Robert L. Wetzel, commander of the V Corps, charged that the West German authorities were not doing enough to crack down on "anarchists and criminals" who had broken into restricted military areas and called for them to be fully prosecuted. His statement led an indignant Interior Minister, Horst Winterstein, to return the critique by accusing the army of having "insufficient security measures at its own installations" and urged the Americans to solve their own problems first before blaming the Germans.[99] In fact, the German police, whose job it was to obstruct the obstructers as it were, went as far as preemptively taking the air out of the tires of vehicles parked near training grounds. Peace activists were prepared, however, and had brought with them replacement tires and spare parts. Nevertheless, some G.I.s did come into direct contact with peace activists. NATO planners, who feared that soldiers may confuse the war games with war and activists with the enemy, gave instructions that G.I.s were to leave the training area as quietly as possible.[100]

When news of the maneuver obstructions trickled back to the United States, many responded with exasperation. Some argued that training exercises in Germany were not that different from similar training exercises carried out in the United States. For example, also in September 1984, 50,000 U.S. soldiers participated in the Operation Gallant Eagle in California. Such maneuvers, however, took place entirely on government-owned reservations, so that the American public was often fully unaware of them, much less inconvenienced.[101] West German Defense Minister Manfred Wörner challenged critical U.S. legislators to "sit in my job for one month" and handle the complaints that pour in. He added: "it is a major American weakness that you are so big, you cannot put yourself in the position of others."[102]

---

[99] *The New York Times*, September 29, 1984.
[100] *Der Spiegel*, October 1, 1984.
[101] *The Washington Post*, September 24, 1984.
[102] Ibid.

Despite opposition from below and disagreements among the elite, the maneuvers continued, year in and year out. At the level of the individual Bundesländer, however, opposition groups were able to prevent maneuvers from taking place by declaring certain areas of land as state parks or protected habitats, and therefore off-limits for maneuvers. On October 11, 1988, the U.S. Forces Liaison Office in Rheinland-Pfalz and Saarland sent a memorandum to the Commander in Chief of the U.S. Army Europe in which the establishment of a nature preserve in Oberes Wiesbachtal was noted.[103] Less than one month later, another memorandum was sent which noted that four separate plots of land in the community of Eppelborn had been declared Protected Landscape Areas and therefore not to be used for maneuvers and training exercises.[104]

Other forms of escalated civil disobedience included the activities of the *Pflugschar* or plowshares group, who were inspired by the biblical admonition to turn "swords into plowshares." On December 4, 1983, shortly after the missile deployment, Wolfgang Sternstein, Carl Kabat, Herwig Jantschik, and Karin Vix broke into the Hardt casern in Schwäbisch Gmünd and began to "disarm" a Pershing II transport vehicle. During the court hearings, prominent members of the peace movement were invited as witnesses, including the theologian Dorothee Sölle, who argued that their actions were a form of "religious observance." In early 1985, the Stuttgart district court announced that the three German citizens would each be fined ninety days' pay, a rather mild sentence considering that they were accused of breaking and entering, property damage, and attempted sabotage. Carl Kabat, a U.S. citizen, did not get off as easily and was sentenced to jail. The judge even described the missile deployment as a "fatal mistake" (*verhängnisvoll*) and said that the missiles should be removed from the Federal Republic.[105] In addition to numerous other plowshare activities, Wolfgang Sternstein also took a leading role in organizing various forms of civil disobedience against the European Command headquarters in Stuttgart throughout 1990s.[106]

Opposition in the Parliament: The Greens

Whereas the Turkish Labor Party was in many ways the initiator of the campaign against the U.S. presence in Turkey, the Green Party in Germany did not play the role of vanguard. On the contrary, issues of peace and security

[103] Memorandum for: Commander in Chief, U.S. Army Europe and Seventh Army from the U.S. Forces Liaison Office, Rheinland-Pfalz and Saarland, October 11, 1988.
[104] Memorandum for: Commander in Chief, U.S. Army Europe and Seventh Army from the U.S. Forces Liaison Office, Rheinland-Pfalz and Saarland, November 7, 1988. The following plots within the community of Eppelborn were declared Protected Landscape Areas: Griesborn, Ill zwischen Wustweiler und Dirmingen, Frankenbachtal, and In Muehborn.
[105] "Vier rüsteten ab mit dem Vorschlaghammer," *Die Zeit*, February 22, 1985.
[106] Interview with Wolfgang Sternstein on May 24, 2006 in Stuttgart. See his autobiography: *Mein Weg zwischen Gewalt und Gewaltfreiheit*, 2004.

did not become a main concern within the Green Party until after the peace movement had become a mass movement. In the original program, only one page out of forty-six was devoted to the question of security.[107] Although some of the regional Green Party offices were critical of military practices such as low-altitude flights already in the late 1970s, as a national party the Greens did not take a position against NATO until 1983, less than a month before the deployment of the cruise missiles.[108] However, after the demand *"Raus aus der NATO"* ("Get Out of NATO") was adopted, it then stayed in the Green Party program until the end of the decade. What exactly this would have entailed, had such a resolution been passed, was never specified. Although it would have likely meant the withdrawal of foreign troops and bases, this was never explicitly stated in any of the party programs. In fact, the various factions within the Green Party often assumed very different positions on the issue.

Some have tried to explain the emergence of the Greens as part of a postmaterialist transformation, arguing that in affluent societies, people become less interested in bread-and-butter issues like wage levels and more interested in issues such as ecology and peace.[109] Such explanations, however, belie the fact that the Greens represented a highly diverse ideological spectrum, ranging from conservative concerns about preserving the German Heimat to leftist critiques of capitalism. Furthermore, many members of the Greens would claim that the concern with the environment and human security is perhaps the most fundamental material concern of all.

Whereas Mehmet Ali Aybar was clearly the leading figure of TIP during the early phase of the party in the 1960s, the German Greens did not have one single charismatic leader of the party during the early 1980s. Instead, there were a number of important founding members of the party, such as Petra Kelly, Rudi Dutschke, Josef Beuys, Eva Quistorp, Roland Vogt, and Gert Bastian. One of the leading theoreticians of the party was Rudolf Bahro. As a dissident in East Germany, he became known as an outspoken critic of the state socialist system; and he was imprisoned in August 1977 after the publication of his book *The Alternative in Eastern Europe*. As a result of pressure from an international campaign, he was released from prison in October 1979 and deported to West Germany. After arriving in the West, it became clear that he was no apologist of capitalism either. According to him, both the East and West suffered from essentially the same system of industrialism, which he believed was not only ecologically unsustainable, but also the root of social problems on both sides of the Iron Curtain. According to Bahro, peace was not defined in the negative sense as an absence of war, but required "a new start with all of civilization.... We must gradually paralyze everything that goes in the old direction: military

[107] Betz 1989: 488.
[108] Betz 1989.
[109] Chandler and Siaroff 1986.

installations and motorways, nuclear power stations and airports."[110] In many ways similar to E. P. Thompson in the UK, Bahro envisioned a type of internationalism from below, composed of peace activists in the West and advocates of democratization in the East, which would undermine the system of military alliances. Bahro's ideas appealed to many of the Greens, as it attempted a holistic approach to peace and security while remaining critical of both superpowers. Furthermore, it incorporated grassroots activism, which was essentially the base from which the Green movement had grown into a national party. Bahro is normally considered as belonging to the "fundamentalist" wing of the Green Party, which is divided between "Radical Ecologists" and "Eco-Socialists." Other leading Green members in this category were Volker Böge and Peter Wilke. For this group, the primary threat to the security of West Germany was not coming from one particular side of the bloc confrontation, but rather stemmed from the antagonism between the superpowers and the alliance system itself.

This position was similar to that of the Neutralists who demanded that the FRG withdraw from NATO. Prominent members of this faction included the previously mentioned General Gert Bastian, as well as Erich Knapp and Wolfgang Schenke who founded the citizen initiative "Peace through Neutrality." On the right-wing of the Green Party spectrum, Alfred Mechtersheimer believed that the problem was not necessarily NATO, but rather the large concentration of both conventional and nuclear weapons on German soil. Because the German government had no control over the use of nuclear weapons, he saw this as an infringement of German sovereignty and demanded the "sovereignization" of Germany.

The left end of the Green spectrum was occupied by the Eco-Socialists. Although other factions of the party were critical of both NATO and the Warsaw Pact, the Eco-Socialists saw the main problem as coming from the United States, which they perceived as an imperialist power. They demanded the demilitarization of the Federal Republic, the withdrawal of all foreign troops and nuclear weapons, and the withdrawal from NATO.[111] Hamburg was the stronghold of the Eco-Socialists, with Thomas Ebermann and Rainer Trampert as the leading figures. The group around Jutta Dittfurth in Frankfurt was another important stronghold of the Fundis.

The Green factions defined the primary threat to Germany as coming either from the presence of nuclear weapons on German soil, from the superpower confrontation (and not solely from the USSR), or from the United States itself. Despite other ideological differences, the Green Party in Germany and the Labor Party in Turkey both rejected the idea that the primary threat came from the Soviet Union.

---

[110] Bahro 1982: 139 ff.
[111] Betz 1989: 491.

Much can be said about the divide between the Fundis and the Realos, but for the current focus on security issues, it can generally be said that the Fundis demanded that the FRG withdraw from NATO, and the Realos thought this was politically impossible. Even if it were politically possible, Realos were skeptical because they saw NATO and the division of Europe as preventing a renewed German hegemony. Whereas the Fundis defended the militant tactics of the peace and anti-nuclear movement, Realos tended to distance themselves from activists who employed civil disobedience or at least did not actively support their efforts.

Of course, such generalizations are only useful to create some order out of the cacophony of details and positions. Inevitably there were exceptions. For example, Otto Schily, who is normally considered as representing the Realo fraction, practiced law and defended Horst Mahler and Gudrun Ensslin, members of the Red Army Faction. Another member of the Green Party but representing the Fundis, Hans-Christian Ströbele, also defended RAF members. In 1970, Ströbele took up the case of RAF leader Andreas Baader. In the late 1980s, a coalition called Green Awakening (*Grüner Aufbruch*) emerged which tried to bring together Realos and Fundis. One member of this coalition, Antje Vollmer, initiated talks with members of the RAF and visited them in prison.

The division between Fundis and Realos was not only a matter of ideological persuasion, but also became embedded in the party structures. Many of the Green parliamentarians in the Bundestag at the time were Realos, but the Federal Executive Board was primarily composed of Fundis.[112] This partly explains why, despite their rhetoric, the Green members of parliament never introduced a resolution to withdraw from NATO. In their 1988 publication "To be or NATO be," the Greens explained that withdrawal from NATO should be understood as a process in which "social forces work towards a politics of substantial disarmament and effective NATO-dissolution" and not a "formal-legalistic act" (*formaljuristischer Akt*). Therefore, no resolution for withdrawal had been introduced in the previous five years, and was unlikely to happen in the near future. The Greens, however, still claimed to be working in that general direction.[113]

Instead, the main parliamentary activity of the Greens consisted of influencing parliamentary debates and obtaining information about security issues through the process of submitting "small and large inquiries" (*kleine und große Anfragen*). These inquiries are traditionally a tool of the opposition parties and are intended as a way of holding the governing party accountable for its actions or, conversely, its failure to act. The inquiries are normally directed to the President of the Bundestag. Small inquiries are usually specific questions which can be answered with the information currently available to the government,

---

[112] Markovits and Gorski 1993: 225–232.
[113] "To be or NATO be: Die NATO-Broschüre der Grünen" published by Die Grünen im Bundestag, Bonn July 1988: 67ff.

and normally a reply is expected within a few weeks. Large inquiries must be answered in writing within six weeks and may require background research. As cited by Zirakzadeh, "[b]etween 1983–87, the Green caucus submitted 367 legislative proposals, 53 bills, 87 inquiries about executive policies and behavior. The Greens also initiated more topical debates than did any other parliamentary group."[114] Just two examples of the types of inquiries that the Green parliamentarians submitted to the Kohl government during the 1980s are (the questions and answers have been translated and paraphrased by the author):

Question: Why doesn't the government explain to the public why the Pershing II missiles are only going to be stationed in the BRD and no other country?

Answer: Because they will replace the Pershing Ia missiles, which are only in the BRD.[115]

Question: How does the government reconcile this with the sovereignty of Germany that it has no control over the use of Pershing II and cruise missiles?

Answer: Germany is not a nuclear power and does not strive to attain this status. The sovereignty of Germany is secured (*gewährt*).[116]

Here we see parallels between the tactics used by the Green Party in Germany and the Labor Party in Turkey, as they both played an important role in exposing information about the U.S. presence. In addition to forcing the government to answer potentially embarrassing questions and obtaining information that may have otherwise been withheld, the Greens also used the court system to oppose the U.S. presence and missile deployment in particular. In 1984, leaders of the Greens filed a suit against the government in the Federal Constitutional Court, alleging that the Pershing II and cruise missiles were designed for offensive use and therefore unconstitutional. For the FRG to be able to accept the missiles, it would have to change the clause of the Basic Law (*Grundgesetz*) that forbids the preparation for war.[117]

The View from Washington

For German domestic politics, the significance of the Greens was manifold. On the one hand, they represented the first attempt to recompose the left after the disintegration of the 1970s into single-issue movements and sectarian splinter organizations. On the other hand, they represented the "march through the institutions" and the greening of the 68ers.

---

[114] Zirakzadeh 2006: 67.
[115] Bundestagsdrucksache 10/249, 14.7.1983, S 14.
[116] Bundestagsdrucksache 10/64 13.5.1983, Frage 86, cited in: Eine Sammlung wichtige Dokumente gegen den Krieg: 39.
[117] The suit was admitted to the court, but the Greens lost the case. Gress 1985: 137.

But what did the Greens signify for the U.S. presence in Germany? For the most part, the Green caucus within the Bundestag was directing its criticism toward the CDU ruling party and its near unconditional support for the U.S. presence. In contrast to the forms of civil disobedience discussed earlier, the Green Party was perhaps only indirectly troublesome for the U.S. military presence. In this sense, the ruling party acted as a "buffer" between the U.S. presence and its opponents in the parliament, just as the large parties in Turkey acted as a buffer between the vituperative critique of the Labor Party and the U.S. forces. And yet despite their relative insulation from the goings-on of the parliament, U.S. military and civilian officials remained highly concerned about the emergence of the Green Party.

In 1980, the Center for Strategic and International Studies of Georgetown University published a report entitled *Ecological Politics: The Rise of the Green Party* by J. F. Pilat. This report was written after the first anti-nuclear demonstrations in Germany at the nuclear power plants of Brokdorf and Whyl in 1976, but before the Green Party had entered the Bundestag and even before the mass demonstration in Bonn in 1981. The report was clearly intended as a first attempt at understanding the rise of the Green Party and the environmental movement more generally. The report hazarded two rather dubious conclusions. First, it concluded that the anti-nuclear protests – which it construed as "antinuclear terror" – could lead to a slowing down of nuclear research and development in the United States. Second, for this reason, it assumed that the Soviet Union was trying to infiltrate or at least influence the antinuclear movement because any disadvantage for the West in the area of research and development would mean a gain for the USSR.[118]

In 1983, the Institute for Foreign Policy Analysis published a report entitled *The Greens of West Germany: Origins, Strategies, and Transatlantic Implications.* By this time, of course, the Washingtonians had had more time to ponder the ramifications of the Greens. The report began, rather immodestly, by claiming responsibility for all of the positive developments not only in Germany but in Western Europe more generally over the past thirty years or so, claiming that "none of the successes registered in Western Europe would have been possible without the security guarantee furnished by the United States in the Atlantic Alliance."[119] However, the fact that the Greens received 5 percent of the vote in the March 1983 elections could – according to the report – potentially undermine the democratic tradition in Germany, and even conjured up the phantom of the weak Weimar Republic:

The announcement of the Greens after the election of March 6, 1983 that they would resort to necessary extra-parliamentary means to prevent the installation of modern NATO missiles brought to mind statements by the Nazis before the fall of the Weimar Republic.[120]

---

[118] Pilat 1980.
[119] Pfaltzgraff 1983: 2.
[120] Ibid: 5.

Ironically, later observers would claim that the emergence of the Greens was proof of the very robustness of German democracy.[121] The question is, of course, whether the author of this particular report and others like it truly feared the subversive potential of the Greens, or if they were merely engaging in polemics or rhetorical denunciations? Because of West Germany's central position in the Cold War, it is indeed possible that political opposition to NATO was taken seriously:

> The realization of the goals of the Greens would detach the Federal Republic of Germany politically, militarily, economically and psychologically from the West. Without the Federal Republic, NATO would be effectively dismantled.[122]

Paul Nitze, Reagan's chief negotiator at the INF Treaty from 1981 to 1984 and a major architect of U.S. Cold War strategy more generally, referred to the Greens in his memoirs as composed of a combination of "radical environmentalists, anarchists, leftists, pacifists, and a few hardened Communists of the Trotskyite persuasion."[123] So as not to lose votes to the Greens, Nitze feared that the SPD would be driven to the left.

Were these fears shared by U.S. liaison officers and the military more generally? For them, the four *Bundesländer* in the American sector were their primary concern: Hessen, Rheinland-Pfalz, Baden-Württemberg, and Bavaria. Of these, Hessen was perhaps most troublesome as it had historically been a stronghold of the Social Democrats and a center of the student movement in the 1960s, whereas Bavaria had been governed by the Christian Democrats and CSU for decades.

After the elections in Hessen on September 26, 1982, the U.S. Forces Liaison Office (USFLO) sent a memorandum to the Deputy Chief of Staff of Host Nation Activities (DCSHNA) reporting on the election results, as was common after both state and national elections. The Greens, who had won 8 percent of the popular vote, were able to enter the Landtag with nine seats. According to the report: "They [the Greens] ran well in metropolitan areas where US troops are located. They are sure to be anti-military and super critical of us when we cause environmental problems."[124] By 1986, the Greens had even entered the state parliament in Bavaria, after winning 7.5 percent of the vote:

> It remains to be seen how their presence in the State Parliament will affect the US Forces. The Forces can reasonably assume, however, that the Greens will miss no opportunity to raise as many inquiries (interpellations) as possible, which will prove time-consuming if not troublesome for the US Forces in general and this office in particular.[125]

---

[121] Hockenos 2008; Thomas 2003.
[122] Pfaltzgraff 1983: 9.
[123] Nitze 1989: 368.
[124] Memorandum to the Deputy Chief of Staff of Host Nation Activities from the U.S. Forces Liaison Office, Hessen, October 5, 1982.
[125] Memorandum from the U.S. Forces Liaison Office, Bavaria, October 16, 1986.

Perhaps more disturbing than the parliamentary debate or outward rhetoric were the increasingly strict environmental standards and the fact that Green Party members were placed in positions of power and able to enforce them. Willy Lehninger, who had served as U.S. Forces Liaison Officer in the state of Hessen from 1982 until 1988 and hence during the height of the anti-base unrest, wrote an end-of-tour summary of the political climate:

> In 1984 the Hessian political scene became one of incipient crisis.... A new Ministry for Environment and Energy was established with a Green Party Minister, Joschka Fischer, and two Green Party State Secretaries were added to the State Government.... The situation had now changed, and a kind of tension existed between the U.S. Forces and the Hessian authorities. The State Government reflected public and official attitudes at local levels which were increasingly critical about our operations. Frictions resulted from the many construction programs we had initiated in past years, and maneuver activities raised public sensitivity. The sensitivity was a result of the growing 'defense weariness' observable not only in Hessen, but also in other parts of the FRG. For instance ... there is no way the continued heavy use of eastern Hessen by V Corps' units can be made fully acceptable to the public.[126]

Opposition on the Bases: Labor Unrest of Civilian Employees

In the mid-1980s, the U.S. Army Europe was the sixteenth largest private employer in the entire Federal Republic, not including the public sector. Because of the geographical concentration of the U.S. Army and Air Force, certain regions were much more dependent on military-related jobs than others. For example, in West Berlin, the U.S. military, together with the British and French, was the third largest employer in the entire city.[127] During the 1980s in Rhineland-Palatinate, the U.S. Forces were the second largest private employer in the state.[128] The only Bundesland that did not have a civilian airport, the Rhineland-Palatinate did however, host eight military air bases, leading Chancellor Kohl to refer to the state as the "aircraft carrier of NATO."

German civilian employees (*Zivilbeschäftigte*) performed a number of jobs that tended to fall under the categories of base support or logistical support. This could include anything from skilled clerical or repair work to unskilled manual labor including maintenance of buildings, mowing the lawn, working in the PX exchanges, and guarding various facilities, among other things. According to Ritzenhofen, 22 percent of the civilian employees worked as skilled or unskilled clerical staff, 12 percent as storage or transportation personnel, 7 percent as mechanics, 6 percent as service or security personnel, and 4 percent as machinists.[129]

[126] Memorandum for Commander in Chief, U.S. Army Europe and Seventh Army from the U.S. Forces Liaison Office, Hessen "End-of-Tour Summary 1982–1988."
[127] "Gute Soldaten, deutsche Kameraden," *Der Spiegel*, January 2, 1984.
[128] Moeller 1995: 144.
[129] Ritzenhofen 1995: 179.

And yet, according to Benton Moeller, who was Chief of the Government Relations Branch at USAREUR headquarters for a number of years, "it was very difficult to find able-bodied Germans willing to work for the U.S. Army in Germany as a whole," which could at least in part be attributed to the "cavalier way" that the U.S. Forces were perceived as treating their employees.[130] As shown later in this chapter, civilian employees who worked for NATO Forces did not enjoy the same rights as employees in other sectors. Most German civilian employees were organized in the Public Service and Transportation Union (ÖTV). The *Deutsche Gewerkschaftsbund* (DGB) is the umbrella organization which represents all individual unions, making it the German counterpart to the AFL-CIO in the United States. There were a number of contentious issues between the U.S. Forces and the trade unions including:

1) co-determination by works councils known as *Mitbestimmung*
2) wage levels/pay scales
3) dependent hire versus the hiring of local citizens
4) the provision of information about plans that would impact stability of employment

Although the issues of dependent hire, wage levels, and the provision of information were areas of concern that affected labor relations between the U.S. Forces and trade unions in any host country, the issue of codetermination was particular to Germany. In contrast to the more adversarial nature of labor relations in the United States, German labor relations were characterized by a higher level of cooperation between employers and unions, even to the extent that employees were involved in management decisions through works councils (*Betriebsräte*). This was known as *Mitbestimmung* or codetermination. Writing in the mid-1990s, Benton Moeller said that regarding works councils, "most U.S. commanders new to Germany are frankly appalled when they see that they will have to share their managerial prerogatives with representatives of the workforce."[131]

The U.S. Forces never entirely recognized workers' rights to codetermination, but instead merely allowed workers to make suggestions, so that employees had a "right to be heard," but not a right to codetermine issues about their work lives. Even after the Federal Representation Law (*Bundespersonalvertretungsgesetz*) was implemented in 1974, the civilian employees still did not enjoy the same rights of codetermination as their colleagues in other sectors.[132] According to Heinrich Linden, who served as Executive Director of the Division of

---

[130] Moeller 1995: 146.
[131] Moeller 1995: 155.
[132] The U.S. Forces agreed to codetermine twenty-three of twenty-eight areas required by the law, with the exception of: hiring, classification to a lower or higher grade, social plan, measures to increase productivity, and the introduction of new work methods. See Ritzenhofen 1995: 178.

Employees of the NATO Forces at the ÖTV union: "no other group of employees has so few rights as the civilian employees of the NATO allies."[133] Although this statement may be an exaggeration, as it can be assumed that undocumented workers in Germany likely experienced harsher forms of exploitation and discrimination, it does attest to the widespread bitterness and sense of being second-class employees.

Despite these frustrations and complaints, the German civilian employees were much less militant than their Turkish counterparts. This can at least in part be attributed to the fact that in general, the German labor movement was more oriented toward corporatism than engaging in class struggle. Nevertheless, many individual union members sympathized with the goals of the peace movement. It is estimated that approximately one-third of the participants (or about 100,000 people) at the first mass demonstration in October 1981 were union members. The unions had paid for approximately 2,000 buses to drive their members from all parts of West Germany to the demonstration in Bonn.[134] While encouraging participation at the protest rally, the DGB did not want to take any leadership role and had forbidden their members from giving public speeches as representatives of the DGB. Georg Benz, who was a member of the executive committee of the metal workers' union (IG Metall), gave a speech at the demonstration; however, he could only speak for himself as an individual citizen, and not as representing his union. He was the only union member who did so, and his actions were admonished by his colleagues.

Despite the participation of rank-and-file union members in the peace movement protests and the rare instances of union leaders giving public speeches, the trade union bureaucracy as a whole was slow to throw its full weight behind the movement. In part, this was attributed to the nature of the administrative apparatus. At the time, the DGB executive (*Bundesvorstand*) was composed of nine members and the heads of the individual unions. It made decisions on a consensual basis, not according to majority rule, so the veto of a single person was enough to prevent resolutions from being passed or to weaken specific statements.[135] Not only did this organizational arrangement make the DGB a weak federation, but the historical experience of German trade unions during the 1930s also contributed to their cautious politicking. During the 1930s, the individual unions were not politically united, but were organized essentially as *Richtungsgewerkschaften*, where some unions had close ties to the Social Democrats and others were closer to the communist party (KPD). According to Reiner Steinweg, many union leaders in the postwar period believed that this division made it easier for the Nazis to crush the labor movement.[136] Hence the DGB was founded with the intention of "never again" being politically divided.

[133] "Gute Soldaten, deutsche Kameraden," *Der Spiegel*, January 2, 1984; see also Linden 1990: 102–103.
[134] Steinweg 1982: 189.
[135] Steinweg 1982: 195.
[136] Steinweg 1982: 195ff.

As understandable as this may be from a historical perspective, it made the day-to-day political work of the organization much more cumbersome. Over time, however, the DGB bureaucracy eventually warmed up to the peace movement. At the June 1982 peace rally, the DGB youth were allowed to participate en bloc. For the Easter March demonstration in 1983, the DGB youth were allowed to publicly call on their members to participate. On October 5, 1983, the DGB organized a five-minute general strike to demonstrate its opposition to the missile deployment.[137] Despite the significance of the event, the American media hardly took notice. *The New York Times* ran a one-sentence blurb and the *Christian Science Monitor* dedicated a full two sentences to the "peace strike."[138] Finally, during the Action Week in late October 1983, the DGB actively took part in organizing the event by making flyers and leaflets and by assigning union leaders to speak as explicit DGB representatives.

Individual unions, of course, had freer rein in terms of taking a stand regarding the missile issue as they only needed to form a consensus within their own union. The IG Metall was and still is the strongest single union in Germany with more than 2.6 million members in the 1980s. In addition to the automobile industry, the IG Metall is also responsible for organizing employees in the defense and shipbuilding industry, which is concentrated in Munich, Kassel, Hamburg, and Kiel. Already in 1978, the IG Metall established a workshop called "*Wehrtechnik und Arbeitsplätze*" to discuss the issue of creating jobs outside of the defense industry. In 1980, the IG Metall demanded that the NATO double-track decision be repealed. By contrast, the ÖTV, in which the base workers were organized, did not pass a resolution that condemned the double-track decision. The IG Metall has traditionally been considered one of the more progressive or militant unions, as many of their campaigns went beyond issues of wage levels to include political demands such as an end to racism and other forms of discrimination. In the mid-1980s the IG Metall won a reduction in the work week to a total of thirty-five hours.

Despite the increased cooperation and coordination between the trade unions and the peace movement, there is no indication in the existing archival material that American officials became troubled in any way. The political implications of the 8 million-member DGB joining the opposition to the missile deployment may have meant that anti-base forces were growing. However, defense planners were less concerned about public opinion, which they already knew was largely opposed to the missile deployment, than they were about the continued functioning of the U.S. protection regime, which was never threatened by labor unrest as in Turkey.

[137] Koch 1983: 108.
[138] *The New York Times*, October 6, 1983 and "West Germany's New Lean toward Missile Deployment," *Christian Science Monitor*, October 5, 1983.

Low-Altitude Flights

Not only was the territory of the Federal Republic subjected to military maneuvers, but also its air space. Both the U.S. Army and the Air Force, as well as the German Luftwaffe and the air forces of other NATO partners used German air space for training purposes. One particularly pernicious aspect of these exercises were the low-level flights known in German as *Tiefflüge*, defined as any flight which takes place at or below 300 meters (900 feet) above the ground.[139] A study by the Starnberger Peace Research Institute regarding low-level altitude flights is revealing: in 1986 a total of 4.6 million flights took place in German airspace, of which approximately 1 million were military flights, including both the Bundesluftwaffe as well as the other Allied air forces. Approximately 100,000 of these military flights were low-altitude flights. When calculated in terms of hours (one flight hour being fifty minutes long) low-level flights were flown for a total of 70,000 hours during 1986. Of these, 25,000 hours were flown by the German air force, whereas the remaining 45,000 were flown by the other NATO partners, in particular the U.S. Air Force.[140] In general, low-altitude flights were only banned over large metropolitan areas, so that two-thirds of the entire territory of the Federal Republic was open for low-level flights. In addition, there were seven areas where military aircraft were allowed to fly as low as 75 meters (250 feet), or treetop level.

Roland Vogt, one of the founding members of the Green Party, headed one of the first anti-low-level flight initiatives in Bad Dürkheim, a resort town (*Kurort*) in the Westpfalz. The area belongs to the wine-growing region of southwestern Germany. During the Cold War, it was part of the U.S. sector; the town of Bad Dürkheim is approximately thirty miles east of Kaiserslautern, which hosts the Ramstein Air Base and the largest overseas U.S. military community in the world. According to Vogt, the constant roar of the military jets was not only linked to higher rates of insomnia and other health problems, but also had a negative economic impact on Bad Dürkheim as a resort town. The proximity to the largest U.S. Air Force base in Europe made it difficult for the small town to promise visitors that they could expect a quiet and relaxing stay. Vogt organized the first public meeting in a restaurant belonging to a local vintner, and was surprised himself that it was attended not only by fellow Greens, but also by locals who traditionally voted for the Social Democrats or Christian Democrats as well. Perhaps even more surprising is that he was able to organize an "anti-flight-day" during which balloons were flown from a nearby hill, which meant that military aircraft would be unable to take off or land. Such comparatively harmless activities could be punished by up to five years in prison if they could be interpreted as "sabotage of

---

[139] By way of comparison, the Empire State Building in New York City is 449 meters tall including the antenna, whereas the cathedral in Cologne is 157 meters tall.

[140] "Tiefflug: Die eigene Bedrohung," Starnberger Impulse, Mediatur Beilage, November 1988.

defense infrastructure" (*Sabotagehandlung an Verteidigungsmitteln*) under
Paragraph 109.

Klaus Vack, the leader of another anti-low-level flight initiative in the state
of Hessen, wrote an article about the growth of such grassroots citizens' initiatives: "These initiatives are being taken just as seriously as parties and other
more established civil society organizations. The politicians take them seriously
because they know that 90% or more of the people who live in the affected
areas are against low-level flights."[141]

Although initiated by the left, opposition to U.S. military activities was
spreading across the political spectrum. Official army documents from Bavaria
provide further evidence of this trend. By the mid-1980s, the Bavarian state
government, known as one of the most conservative in the FRG, was demanding that the Federal Ministry of Defense should redistribute low-flying aircraft
to other parts of the country because the training operations in Bavaria had a
negative impact on tourism and public health. A document from the USFLO
office in Bavaria describes the situation as follows:

> The lesson here is that this discussion took place between political allies who have
> the same outlook on defense-related matters. It was not a confrontation between the
> Defense Ministry and a group of local radical malcontents.... in the prevailing mood of
> the 1980s, local interests and sensitivities are more likely to be recognized than vetoed
> regardless of basic orientations and party affiliations.[142]

The state parliament (*Landtag*) of Rheinland-Palatinate passed a resolution on
July 6, 1988 to reduce aircraft noise and prohibit low-level flights. The parliament demanded the development of defense alternatives that would make
low-level exercises unnecessary, including the development of computer simulations. Pilots who violated these regulations should be punished. The decision
represented a common resolution of all four party fractions (CDU, FDP, SPD,
and Greens) and was passed unanimously. As noted in a memorandum written
by the U.S. Forces Liaison officer in Rheinland-Pfalz, this was the first time that
a state parliament had passed such a resolution.[143] Furthermore, it should be
emphasized that this was still before the catastrophic plane crash in Ramstein
in August 1988.

Military Air Shows and the Disaster at Ramstein

For several decades, the U.S. Air Force had been hosting an annual flight show
at the Ramstein Air Base at which visitors were allowed onto the military base
while pilots from various NATO countries displayed daring flight acrobatics.

[141] "Der Fluch mit dem Tiefflug – oder: Widerstand ist möglich" in: "Tiefflug: Die eigene
Bedrohung," Starnberger Impulse, Mediatur Beilage, November 1988.
[142] Memorandum from USFLO Bavaria, March 1, 1985.
[143] Memo for Commander in Chief, U.S. Army Europe and Seventh Army from U.S. Forces Liaison
office, Rheinland-Pfalz and Saarland on July 11, 1988.

During the flight show on August 28, 1988, Italian fliers collided in mid-air, killing seventy people and injuring many others. According to press reports, the Ramstein air disaster stirred "unprecedented debate" over allied military activities. One U.S. official who wished to remain anonymous went on record as saying: "Ramstein has become a catalyst for opposition to a variety of allied operations in Germany."[144] Immediately after the disaster, Defense Minister Rupert Scholz announced a permanent ban on aerobatics in air shows which applied to both the Bundesluftwaffe and all allied forces stationed in Germany. However, questions arose about whether he had the authority to do this. For example, the Standardization Agreement (Stanag) 35/33 did not regulate *if* the U.S. and other allies were allowed to host flight shows, but rather *when* they did, which safety measures had to be in place.[145] U.S. officials voluntarily cancelled an air show scheduled for September 4 and generally remained quiet about the legal situation, trying to avoid opening a debate about allied rights in the Federal Republic. When pressed to make public statements, U.S. officials tried to distinguish between stunt flying and low-level flights. Although stunt flying was intended to garner support for the NATO presence and hence served public relations purposes, officials continued to claim that low-level flights were necessary to maintain combat readiness. However, the public increasingly saw both stunt flying and low-altitude flights as dangerous and as a violation of German sovereignty.

In fact, among both legal experts and the opposition parties, a debate had already begun about the issue of the U.S. and NATO military presences and to what extent it impinged on German sovereignty. For example, in April 1988 an article appeared in the *Neuen Juristischen Wochenschrift* about the fact that the U.S. military applied the death penalty to its own military personnel. Between 1982 and 1988, four U.S. soldiers had been sentenced to death in Germany, although in the Federal Republic the death penalty was officially banned. After the Ramstein accident, even high-ranking officials such as Rudolf Sharping, the head of the SPD in Rheinland-Pfalz, began to demand an end to the "dictates" of the allied militaries.[146]

The Ramstein accident was neither the first nor the last military aircraft accident in 1988. On December 8, 1988, a U.S. Air Force A10 crashed in a densely populated neighborhood in Remscheid, killing the pilot and five residents. A month later, on January 13, 1989, two West German air force jets and a British warplane collided during low-altitude training over northern West Germany.[147] In 1988 alone, eighteen NATO aircraft crashed in West Germany, one of which fell close to a nuclear power plant. By this point opposition to

---

[144] *Christian Science Monitor*, September 19, 1988.
[145] "'Eine Art Besatzungsrecht': Das NATO-Truppenstatut schränkt die Souveränität der Bundesrepublik ein," *Der Spiegel*, September 12, 1988.
[146] Ibid.
[147] *The Washington Post*, February 2, 1989.

low-altitude flights had spread from the affected areas where low-altitude flights were allowed to encompass citizens from all over the Federal Republic. According to a survey by the television station ZDF, 71 percent of the entire population was in favor of banning low-altitude flights. On October 6, 1988, the Darmstadt Administrative Court (*Verwaltungsgericht*) ruled that low-level exercise flights below 300 meters should be banned. The court based its findings on the fact that it was "not convinced that the carrying out of low-level flights under 300 m is necessary for the maintenance of defense capabilities." In the Netherlands, also a NATO partner, it was illegal for military aircraft to fly below 300 meters.[148] On December 15, 1988, the governors of all ten West German states plus the mayor of Berlin agreed unanimously to seek a reduction in low-level flights.[149] The table below gives an overview of the 203 military plane crashes that occurred in less than eight years. Although meant to reassure allies and create protection, military training exercises may not inevitably, but often enough have instead resulted in the production of what I have termed collateral harm.

Number of military plane crashes in individual Bundesländer between 1980 and October 1988	
Baden-Württemberg	26
Bayern	46
Bremen	1
Hamburg	0
Hessen	16
Niedersachsen	37
Nordrhein-Westfalen	34
Rheinland-Pfalz	31
Saarland	5
Schleswig-Holstein	7
FRG Total	203

In September 1988, less than one month after the Ramstein catastrophe, the fall REFORGER maneuvers began. The September 1988 REFORGER involved 125,000 ground troops, 103,000 of whom were Americans, and 30,000 vehicles, 7,000 of which were tracked vehicles such as tanks and missile launchers. The maneuver took place over 16,000 acres of central and southern Germany. This turned out to be the most disastrous military training exercise ever in the history of the FRG. Seven people lost their lives, including five German

[148] Memo for: Commander in Chief, U.S. Army Europe and Seventh Army, from the U.S. Forces Liaison Office, Hessen on October 24, 1988.
[149] *The Washington Post*, January 2, 1989.

civilians. Public opposition to these massive maneuvers became so widespread that the 1989 REFORGER exercise was cancelled.

As most social movement victories are short-lived, in 1990 the REFORGER exercises took place again. However, the number of troops involved was reduced by half, the use of tanks was prohibited, and computer simulations replaced many of the previous training exercises. A report in *Spearhead Magazine* describes the changes in the 1990 maneuvers:

This REFORGER was quite different from the previous 21 exercises that have been held almost annually for the past two decades. This exercise involved fewer troops, fewer tracks, no tanks and lots of computers. The total number of participants for this REFORGER was 55,000 troops; compared with the 97,000 soldiers that took part in the 1988 REFORGER, troop involvement this year has been cut by 47 percent.[150]

Conclusion: From Protective Power to Pernicious Protection

To return to the conceptual issue of protection regimes, I aimed to show that the nature of the U.S. presence in Germany changed dramatically over time. After beginning as an occupying power, by the time the Berlin Airlift was over in the late 1940s, and before the official occupation regime had even ended, the U.S. military had established itself as a benevolent hegemon offering protection from the Soviet Union. Germans used the word "*Schutzmacht*" to refer to the U.S. presence as a "protective power." By the late 1970s and especially by the early 1980s, however, the U.S. protection regime in West Germany was rarely referred to as a *Schutzmacht*, but was approaching the other ideal type, what I have called a pernicious protection regime. This change was particularly clear among those living in the American sector who were subjected to the constant firing of the shooting ranges, the roar of the low-altitude flights, and the widespread damage caused by the incessant military exercises. After Ronald Reagan began talking about the possibility of winning a "limited nuclear war" which would have only been limited for Americans, but which would have meant serious destruction of the European continent, voices critical of the U.S. military presence could be heard throughout the Federal Republic. A quote from the editorial of the *Blätter für deutsche und internationale Politik* from May 1981 is perhaps the most poignant expression of this change in the protection regime:

What type of "protection" is this, which is bought at the price of an increased danger of war for us, the protégés? What type of a "protective power" (*Schutzmacht*) is this, for whose Secretary of State there are more important things than to live in peace, while for our country war would mean certain demise? What kind of a "protective power" is this, which plays with the idea of a "limited nuclear war" … while they reduce the risk to themselves by taking Western Europeans as their "atomic hostages?"[151]

[150] "REFORGER 1990: Centurion Shield" by K. M. Shimko in *Spearhead Magazine*, January 1990.
[151] Editorial by Karl Bredthauer in *Die Blätter*, Vol. 5, No. 26, May 1981: 521.

Of course, other observers have already described the rise in a certain anti-American sentiment in Europe during the early 1980s. However, so far none have examined to what extent this unrest interfaced with or had an impact on the U.S. presence. In public, Pentagon officials have generally been reluctant to admit that any policy changes are a result of pressure from public opinion or social movements. For example, after the large maneuvers were scaled back, defense officials claimed that this was attributed to "improved training techniques."[152] The official military records, however, would tell a different story.

In document after document, it became clear that protest events were rarely dismissed as trivial or tangential. When U.S. commanders planned to move to northern Germany in the second half of the 1970s and establish a new base in Garlstedt, although it was not in the U.S. sector, a Community Relations Advisory Council (CRAC) was formed and the first year of planning for the stationing program included "extensive analysis to place into proper perspective the various protest groups."[153] After observing the local scene, including the emergence of two citizens initiatives, one of which presented a list of 45,000 signatures of citizens opposed to the new base, it was determined that "[t]he primary challenge existed in the form of small vocal protest groups opposing the concept and possible consequences of stationing US Forces in the NORTHAG area."[154] The primary concern of the citizen initiatives was the environmental impact of the new facility. The Lower Saxony legislature approved of the stationing only if the rights and concerns of the local citizens were "safeguarded to the greatest extent possible." USAREUR understood this as a "warning." Hence, the analysis concluded that: "It had to be recognized that the protest groups posed a problem for the national level staff in their efforts to convince local and state government officials to accommodate the U.S.-F.R.G. agreement."[155] The way in which the United States responded to the opposition was instructional: rather than dealing directly with the protest groups, the United States deferred to the German authorities. In fact, no attempt was made to convince the protest groups that the U.S. expansion into the British sector of northern Germany made sense. However, the army launched a public relations campaign which addressed many of the concerns raised by the citizens initiatives. The army presented itself as environmentally friendly, by publicly announcing the removal of thirty tons of dud ammunition from the heath before beginning the construction. It also endeavored to present the U.S. soldiers not as members of a foreign military force, but rather as civilian ambassadors or exchange students, and launched the New Neighbor Program in which individual soldiers

---

[152] "U.S. Decides to Scale Back Military Exercise in Europe," *The New York Times*, October 25, 1989.
[153] "Strengthening NATO: Stationing of the 2nd Armored Division (Forward) in Northern Germany" published by the Headquarters, United States Army, Europe and Seventh Army, May 22, 1980.
[154] Ibid: 52.
[155] Ibid.

would be paired with German "host families." In sum, although the opposition was not able to prevent the building of the Garstedt base, it succeeded in making the U.S. Army respond to its concerns.

In the Historical Review published by the USAREUR History Office covering the years 1982 and 1983, protest activities were discussed at length. In contrast to much of the academic literature on this period, which has classified the social unrest as a strictly anti-nuclear movement or peace movement which the German government and police forces had to deal with, official documents from the U.S. Army confirm my claim that this protest activity is perhaps better understood as anti-base unrest which posed a challenge for the American military presence as well:

While the central demonstration theme appeared to focus on impending deployment of the Pershing II and ground-launched cruise missiles (GLCM), all US forces activities, installations, and members were potential targets of deliberately planned activity, or as targets of opportunity.[156]

Furthermore, there were weekly "demonstration forecast updates" that were provided by the U.S. Army Europe to U.S. force members, who received detailed instructions on how to react if they personally encountered protest events.

In 1983, a report was written entitled "The German Context within which USAREUR must live and operate." Part E of the report summarizes the changed political context and makes clear that the army was experiencing problems on all sides. Those parts of the American public and U.S. Congress which had become cognizant of the protest movements in Europe, began to question whether the large U.S. presence in the FRG made sense anymore. The USAREUR could also no longer rely on their "political friends" in Germany as much as they could in the past, and according to the report "everything we do or plan to do, has become a political act." There are six main lessons that the report outlines regarding how to operate and respond to unrest in the political environment of the 1980s:

1. Convince Congressional visitors that the Germans are responsible and that the Germans are doing their share for NATO.
2. Make sure that the German government is responsible for dealing with opposition to the U.S. Army, it is not the responsibility of commanders to explain to the Germans why they are here.
3. Don't ask too much of our German political friends.
4. Spread news about how the German economy is positively affected by our presence.
5. Consider that it is necessary to communicate our plans in a timely manner.
6. Plan and operate in a sensitive manner.[157]

---

[156] Historical Review 1985: 133.
[157] "The German Context within which USAREUR must live and operate," U.S. Army Europe and Seventh Army, Heidelberg 1983.

A year later, the Deputy Chief of Staff for Host Nation Activities, Colonel Fredrick C. Schleusing, thought he had found a solution to the litany of problems with which the U.S. military was confronted. In June 1984, a report was prepared on the "Financial and Employment Impact of the US Forces on the German Economy in Fiscal Year 1983." The report laid out several examples of how it had become increasingly difficult to "operate and train in Germany." Three of these were:

1. German local governments, commercial firms and residential developers increasingly find our presence to be a barrier to civil economic development plans and are encroaching on our facilities.
2. Complaints about our environmental offenses due to antiquated heating plants, old and inadequate sewage and waste disposal systems, aircraft and weapons range firing noise have become legion.
3. In recent years, many municipalities have objected to the density of our troop levels and complain about what they call sociological overload [*sic*].[158]

According to the report, these problems derived not from the U.S. military or the way it operated, but rather because some parts of Germany were "dangerously overdeveloped." Colonel Schleusing believed that the best way to deal with these problems was to stress the positive. In his words, "we should turn from step to step retreat and take a positive approach." This meant emphasizing the fact that the U.S. Forces employed 56,130 local nationals, most of whom were German citizens. In addition, the modernization of the U.S. Forces, including the multiple construction projects, even generated extra employment for local citizens. The unemployment rate in Germany had gone from 4.8 percent in 1980 to 9.5 percent at the end of 1983. During this same period, the U.S. Forces had employed an additional 6,000 people. The report even claimed that the U.S. military may have been the only "contra-cyclical" influence on the German economy which had recently gone into a recession. For this reason, the report was optimistic in concluding that: "It may be possible to convince enough influential people that the benefits to the local population of our presence greatly outweigh the life quality costs of our presence."[159]

By the end of the decade, however, the situation had not improved in any significant way. Michael Hanpeter, U.S. Forces Liaison Officer in Bavaria from 1981 to 1988, argued that the liaison officers should be placed directly under the Chief of Staff so as to better deal with the "progressively more difficult environment in which the US Forces are operating in the Federal Republic of Germany in the 1980s."[160]

[158] "Financial and Employment Impact of the US Forces on the German Economy in Fiscal Year 1983," report authenticated by Colonel Fredrick C. Schleusing, DCSHNA on June 5, 1984.
[159] Ibid.
[160] Memorandum from USFLO Bavaria on August 26, 1987.

Willy Lehninger, who had served as U.S. Forces Liaison Officer in the state of Hessen from 1982 until 1988, wrote an end-of-tour summary of the political climate, part of which was cited earlier regarding the tense situation between the U.S. forces and the Hessian authorities after the Greens entered the Bundestag. In the same report, Lehninger continues to describe what he sees as a subtle change in the political stance of the federal government in Bonn, although the Kohl Administration officially maintained its pro-Atlantic position:

It is increasingly apparent that local governments have learned to resort to the court system to stop or delay projects they oppose. This is a source of much concern to me, particularly since the Federal Government in many cases is reluctant to provide us the strong support we need to accomplish our force modernization.[161]

Wiesbaden and Wildflecken were two primary examples in which court cases had put a halt to U.S. plans. In a letter from Defense Secretary Weinberger addressed to "Dear Manfred," the German Minister of Defense on October 16, 1987, Weinberger reminds his counterpart that the modernization program which began in 1980 has been the "most ambitious" in its history and has included $20 billion for conventional force modernization.

Unfortunately, however, many of the advantages of these modernized systems may be undercut through litigation as environmental groups in the FRG become more organized in their opposition to the activities of our forces.[162]

According to other army officials, the court cases were merely an expression of a political sea change. It was not legal issues, but rather "political pressures and environmental awareness" which had "taken over in determining how we can train, for how long and under what conditions."[163]

To return to our original question, how did social unrest in Germany impact the American military presence in its most important host nation? The anti-nuclear movement had clearly failed to prevent the missile deployment. After the peace movement had reached its peak in the large metropolitan areas in 1982 and 1983, it began to decline in the following years. Around the same time, the Red Army Faction had become entirely discredited even among those who espoused violent tactics, especially after they had murdered Edward Pimental in 1985, an enlisted G.I., to obtain his uniform and ID. However, nonviolent anti-base networks continued to grow in both urban and rural areas and remained resilient over the years, posing a sustained challenge to the American military presence.

Defense planners believed that low-level flights and military training exercises were necessary to maintain combat readiness, and were meant to reassure

---

[161] Memorandum for Commander in Chief, U.S. Army Europe and Seventh Army from the U.S. Forces Liaison Office, Hessen "End-of-Tour Summary 1982–1988."

[162] Letter from the Secretary of Defense Caspar Weinberger to Minister of Defense Manfred Wörner on October 16, 1987.

[163] Memorandum from USFLO Bavaria on August 26, 1986.

allies that NATO was able, in the event of an attack, to inflict intentional, concrete harm on Warsaw Pact countries. Increasingly, however, the production of security resulted in unintentional, collateral harm such as military plane crashes or accidents caused during war games. The U.S. protection regime disenfranchised, and at times even harmed the very people it claimed to protect. Anti-base activists were therefore motivated less by ideology than by what they saw as the necessity of limiting the very real harm caused by U.S. military activities.

By the end of the 1980s, instead of appearing self-confident about emerging as the victor of the Cold War, the U.S. military felt that it had been placed in a straightjacket: it was unable to carry out the Master Restationing Plan, unable to continue large-scale maneuvers, unable to upgrade the firing range at Wildflecken or station helicopters in Wiesbaden, unable to even introduce the idea of the "fort concept" because of the political environment, much less implement it. Finally and perhaps most tellingly, U.S. military officials were warned against attending public events where the composition of the audience could not be determined in advance. For many inhabitants of the American sector, the U.S. presence was no longer seen as a *Schutzmacht*, but as more threatening than the threat against which it claimed to protect, and hence becoming what I have termed a pernicious protection regime. The social unrest described in this chapter did not result in the U.S. protection regime becoming dysfunctional, as was the case in Turkey in the late 1960s and again from 1975 to 1978. The question of understanding these different outcomes will be taken up in the next chapter. Despite the fact that there was no breakdown in the protection regime as in Turkey, the various anti-base initiatives did succeed in decreasing the autonomy of the U.S. military commanders, even before the Berlin Wall became history.

4

From Shield to Sword

The End of the Cold War to the Invasion of Iraq

INTRODUCTION

Having outlined the historical trajectory of the U.S. presence in Germany and Turkey during the Cold War and analyzed the limits to the autonomy of the U.S. military during this period, it is now time to turn our focus to the post–Cold War period. With the collapse of the Soviet Union, many things had changed. The end of the USSR represented both a victory and a conundrum for the Defense Department. It was a victory not just in an ideological sense, but there were territorial gains to be made as well: structural limits to U.S. military expansion, embodied perhaps most perfectly in the Berlin Wall, were no longer in place and the post-Soviet space was available for penetration. However, with no clear enemy, it became difficult to justify the continued existence of overseas bases as providing protection against a no-longer existing threat, much less the necessity of expansion. Although the U.S. may have "won" the Cold War, it had lost much of the credibility necessary to maintain a legitimate protection regime. With no external threat, the second criteria necessary for legitimate protection was also lacking.

Not only had the political landscape changed in which the U.S. presence was embedded, but the U.S. military itself had changed as well. During the early Cold War, U.S. troops, bases, and technology were seen by many as the harbingers of modernity and Westernization, for both Germany and Turkey. Now, the presence was a relic of a bygone era. Instead of withdrawing from Germany after the end of the Cold War, as the Russian troops had done, the Americans (along with a few of their NATO allies in much smaller numbers) decided to stay.

Over the years, many U.S. military personnel had grown accustomed to the first-rate infrastructure in Germany, including the civilian support facilities such as the U.S.-owned and operated golf courses and luxury resorts located in the Bavarian Alps. But the reason the Americans were staying was not just a matter of sentimentality or slow bureaucracy. They had plans for the future as well. Instead of functioning as a defensive shield against an outside threat, the U.S. presence was transformed into a sword that could engage in wars of aggression. The Iraq War, which began in 2003, was one of the first examples of the new strategy. As many observers have pointed out, this was the first time that the bases in Turkey and Germany were to be used for a preemptive attack in a region of the world that was clearly outside of NATO's area of responsibility. Whereas the First Gulf War had been sanctioned by the UN, the war in 2003 lacked both UN approval and any legitimacy in terms of international law. What fewer observers have described, however, is how the U.S. presence itself had been transformed. The dual nature of threat and protection, which the U.S. military presence had embodied throughout the Cold War, took on a new urgency in the context of an actual invasion. Rather than representing a comforting "security umbrella" against the Soviet Union, which NATO members had for years considered a common enemy, an invasion was now being planned against Iraq – a country which had not threatened the United States, Germany, Turkey, or any other member of the NATO alliance. The execution of concrete harm was no longer a mere possibility, but a reality. In the context of war, the language of international harm may appear euphemistic. The distinction, however, between abstract, unintentional harm and concrete, intentional harm is still very relevant. At the time of writing, it is estimated that more than 111,903 Iraqi civilians have been killed since the invasion in March 2003.[1] The human suffering caused by the war is immense. Numerous books, articles, blogs, and documentary films have documented the catastrophic suffering caused by the war. In keeping with the overall goal of this book, however, this chapter does not document the misery and destruction caused by the U.S.-led invasion. Instead, it analyzes the social movements that tried to prevent the invasion from happening.

As a number of observers have pointed out, the Iraq War triggered a generalized crisis between Europe and the United States, with some even declaring the "end of the West" while other more cautious observers predicted a major transformation of the Atlantic order.[2] In addition to disagreeing over the extent or severity of the problem, scholars have also disagreed over the exact nature of the conflict. Perhaps most famously, Robert Kagan has argued that the conflict is a result of power disparities between the United States and Europe.[3] Others believe that different interpretations of international law or changing

---

[1] As of April 11, 2013, the number of civilians who have been killed in Iraq is estimated between 111,903 and 122,408. Iraq Body Count, http://www.iraqbodycount.org/ Last accessed April 11, 2013.

[2] See *The End of the West? Crisis and Change in the Atlantic Order*, edited by Anderson, Ikenberry, and Risse for a collection of essays on this topic.

[3] Kagan 2003.

values have led to the discord.[4] While not making any judgments regarding the larger crisis within NATO, by focusing on two key European allies, this chapter contributes to the broader debate. Power disparities between the United States and Europe did not emerge in 2003, but have characterized the relationship for decades. In fact, power disparities were in many ways much larger during the Cold War, in particular in the early period when Europe was still recovering from the devastation of World War II. Furthermore, European and American values have never been identical; hence, it would seem that changing values or power disparities cannot explain what happened in 2003. One thing that has changed significantly, however, is the nature and function of the U.S. military presence in Europe. Strangely, this has been largely missing from the discussion thus far. What I hope to illustrate in this chapter is that the crises with Germany and Turkey were both fundamentally related to issues involving the U.S. military presence.

At the level of bilateral relations, the episode in 2003 led to the worst crisis in U.S.-German relations since the early 1980s, and the worst crisis in U.S.-Turkish relations since the turbulent decade from the mid-1960s to the mid-1970s. As during the Cold War, the crises in 2003 were fundamentally related to issues involving the U.S. military presence. It is important to point out that this was not the case in the dispute with the French, as there were no longer any U.S. bases in France which could have been used for the invasion. Hence, there was much more at stake in the relationships with Germany and Turkey and the crises were much more profound. Furthermore, the crises were characterized not only by elite disputes, but also by widespread mobilization from below. It is necessary, therefore, to take a closer look at these conflicts and to what extent pressure from below factored into the political process. In doing so, we are in a better position to understand the social forces which have led to a questioning or undermining of the political order and for what reasons.

The vast majority of the population in both countries was opposed to the Iraq War.[5] Major social movements emerged in both Germany and Turkey, which, in terms of sheer numbers, resulted in larger anti-war demonstrations than during the Cold War. The governing parties in both countries were also both at least verbally opposed to the war and both had the ability to deny basing access. And yet, despite rhetoric to the contrary and a great deal of blustering, Berlin opened up its airspace and consented to the provision of basing access, and thus enabled the Iraq War. Ankara, however, did not cooperate,

---

[4] Jürgen Habermas argues that the "normative authority" of the United States "lies in ruins" and that it is necessary to build a European identity in opposition to the United States; see Habermas 2006. Although he sees disputes over international law as being at the heart of the Iraq crisis, Byers still believes that the United States and Europe continue to share a common understanding of international rules and norms, Byers 2008. For a less optimistic assessment, see Bartholomew 2006.

[5] For public opinion in Germany, see Haumann and Petersen 2004; for Turkey see Uslu, Toprak, Dalmis, and Aydin 2005.

although it had many more incentives to do so as it was offered large financial aid packages as well as political support for the EU accession process. By refusing to provide basing access, Ankara made it impossible to open a northern front for the invasion. Despite social mobilization from below and an allegedly hostile governing coalition, the U.S. protection regime in Germany remained intact and highly functional, as virtually every access request was granted. In Turkey, however, mobilization from below created enough pressure on the AKP government that it voted to deny basing access. The military infrastructure in Turkey, which had been upgraded and prepared for the war just weeks earlier in expectation that the resolution would pass, was suddenly rendered dysfunctional. What can account for the difference in outcome?

In what follows, I first briefly discuss changes in the U.S. overseas presence since the collapse of the Soviet Union and the attacks of September 2001. I then outline the preparations for the Iraq War and how the Bush Administration envisioned that the U.S. military infrastructure in Germany and Turkey would be put to use for the invasion. I then analyze the positions of the governing parties as well as the opposition movements in both countries. Finally, I return to the question of the changing overseas presence during and after the initial invasion in 2003, including the controversy surrounding the Overseas Basing Commission, and what the history of the U.S. presence in Turkey and Germany can teach us about how the global network of bases may expand or contract in the future.

THE 2001 QUADRENNIAL DEFENSE REVIEW

Shortly after becoming Secretary of Defense, Donald Rumsfeld undertook a top-down review of the overseas force structure, while at the same time the Quadrennial Defense Review (QDR) was being prepared. The QDR, which had been completed before the attacks in New York and Washington on 9/11, was published on September 30, 2001 and announced the need for major changes:

> During the latter half of the 20th century, the United States developed a global system of overseas military bases primarily to contain aggression by the Soviet Union. U.S. overseas presence aligned closely with U.S. interests and likely threats to those interests. However, this overseas presence posture, concentrated in Western Europe and Northeast Asia, is inadequate for the new strategic environment, in which U.S. interests are global and potential threats in other areas of the world are emerging.[6]

In short, the QDR was advocating further expansion of America's baseworld. As significant as the attacks on 9/11 were, they were not the cause for the major realignment of the U.S. overseas military presence, because this transformation was not only planned but was already underway before September 11, 2001. In fact, the 2001 QDR was merely post-factum stating what had already become

---

[6] Quadrennial Defense Review 2001: 25.

reality: the United States had established itself in areas previously outside its sphere of influence and partially closed down the Cold War infrastructure in Western Europe and Asia. The attacks in Washington and New York then gave momentum to a trend that was already clear. After the end of the Cold War the United States has either established new bases, acquired basing access rights, or expanded its already-existing presence in Kuwait, Oman, Bahrain, Qatar, Yemen, Djibouti, Kosovo, Bosnia, Albania, Hungary, the Czech Republic, Poland, Uzbekistan, and Afghanistan.[7] In short, the Pentagon was shifting its energies and resources to the Middle East, Central Asia, the Balkans, and Africa while Western Europe was becoming a quiet hinterland. This meant that Germany was no longer at the front lines of the confrontation, whereas Turkey was increasing in strategic importance, as had already become clear during the Gulf War in 1991. So confident was the United States of its status that it no longer needed to hide behind client states to exercise its influence.

Instead of relying on a local power to serve as a surrogate enforcer of U.S. Gulf policy, as it had done with the shah's Iran before the revolution of 1979 and with Saddam Husayn's Iraq during the Iran-Iraq War, the United States appointed itself as the protector of the Gulf. U.S. ships patrolled the waters east of Suez, its warplanes made daily surveillance flights over Iraq, and its troops remained permanently stationed in the region.[8]

In addition to acting as "protector" of Europe and East Asia, the United States was now also "protector of the Gulf." The nature and size of the U.S. presence in these different regions of the world varies considerably. In general, however, the Pentagon has tried to emphasize that it is no longer interested in establishing a permanent presence overseas, but instead wants permanent *access* to military facilities. In some cases, this means using already existing facilities, whereas in other areas it means establishing new bases. Defense officials have created new terms such as cooperative security locations (CSLs) or forward operating locations (FOLs) to refer to bases that are meant to be austere, inconspicuous facilities. Some have even proffered the term "lily pads." As Michael O'Hanlon has pointed out, however, there is an inherent contradiction within the basing realignment plan: on the one hand, it claims to simultaneously reduce the overseas presence while also being better able to fight terrorism by getting closer to the terrorists, who are said to inhabit the "arc of instability" spanning large parts of Africa, the Middle East, and Central Asia.[9] This means of course establishing more military installations in more regions of the world. In other words, whereas the number of large garrison towns may decrease, the number of military bases which the United States has access to will increase.

---

[7] For a description of the increased U.S. military presence in the Middle East and Central Asia see Lawson 2004, Johnson 2004, and Johnson 2006.
[8] Cleveland 2000: 501.
[9] O'Hanlon 2008.

The implications of this transformation will be discussed further at the end of this chapter, but for long-standing NATO allies the general changes are clear. The primary purpose of the U.S. military presence in Germany and Turkey is no longer to provide protection to Germany or Turkey. But instead, the U.S. military is mainly interested in maintaining its bases NATO countries so that they can be used to project power elsewhere, as potential launch pads for invasions.

GERMANY

As indicated previously, the changes in the overseas basing network in the post–Cold War era involved both expansion and contraction. Later in this chapter, I briefly describe how Germany was impacted. During the second half of the twentieth century, the Federal Republic of Germany hosted the largest number of U.S. troops of any country in the world: nearly three-fourths of the European-based forces were stationed there, or approximately 250,000 personnel. Since the end of the Cold War, there have been two major draw-downs of the U.S. presence in Germany. The first draw-down happened during the 1990s.[10] In 1989, the U.S. military presence in Germany consisted of 213,000 soldiers along with 65,000 civilians stationed at approximately 858 installations. Ten years later, there were 62,000 soldiers and 17,000 civilians located at 265 installations in Germany.[11] Although shrinkage was the main tendency throughout the 1990s, with many of the smaller army posts along the former border to East Germany being closed, some air force bases such as Ramstein and Spangdahlem were expanding. Consolidation around major hubs was an important part of the transformation process (Maps 4.1 and 4.2).[12]

Perhaps the most symbolic and most reported-about feature of the second draw-down was the return of the Rhein-Main Air Base to the Frankfurt Airport and the subsequent expansion of Ramstein. For decades, Frankfurt had been the port-of-entry and "Gateway to Europe" for American soldiers and their families.[13] As noted during the farewell ceremony, more than 15 million Americans had lived in Germany over the past sixty years. Although the closure of Rhein-Main was completed by the end of 2005, further reductions in the number of installations and troops were scheduled to continue throughout the decade. This second major draw-down was to be part of a larger transformation of the overseas presence globally, which then-Governor Bush had made a campaign issue during the 2000 Presidential election.

After the attacks of 9/11, Chancellor Schröder declared "unconditional solidarity" with the United States as large crowds gathered in front of the American

---

[10] For an analysis of the drawdown in the early 1990s, see Cunningham and Klemmer 1995.

[11] Bryan van Sweringen: "Stationing within the State: The U.S. Army Presence in the Rhineland-Palatinate 1947–2007," paper presented at the conference "Amerikaner in Rheinland-Pfalz" in September 2007.

[12] Cunningham and Klemmer: 1995.

[13] "Abschied von Good old Germany," www.dradio.de/dif/sendungen/hintergrundpolitik/453231.

MAP 4.1. Map of US Presence in Germany 1990.

embassy in Berlin to lay flowers and wreaths. A majority of Germans polled even agreed that Bundeswehr soldiers should participate in counter-measures within NATO.[14] Schröder supported the mission in Afghanistan, surviving a

[14] Noelle-Neumann and Köcher 2002; see also Forsa, Meinungen zum Verhalten der Bundesrepublik gegenüber den USA bei einem Irak-Krieg, survey published November 30, 2002, cited in Szabo 2004: 83.

MAP 4.2. Map of US Presence in Germany 1996.

no-confidence vote when members of his own SPD and the Greens opposed the deployment of Bundeswehr soldiers to Afghanistan.[15]

When the war-planners in the Bush Administration began drawing up plans for the invasion of Iraq, they looked to their allies in Berlin and Ankara. To

---

[15] Katzenstein: 2002.

open up a northern front for the invasion of Iraq, the use of Incirlik as well as other facilities in Turkey was considered crucial, and the bases in Germany were equally critical as a staging ground for the invasion. Despite the fact that the Department of Defense had acquired bases in more than a dozen countries, the massive infrastructure in Germany had no equal; and the geographical location of Turkey, which shares a border with Iraq, made it indispensable as well. Although over the past decade the network of overseas bases has expanded globally, the United States is still highly dependent on a few key allies.

The Iraq War

For the invasion, defense planners counted on unimpeded access to the basing infrastructure in Turkey and Germany, including use of the airspace. The facilities in Germany were among the most expensive real estate that the Pentagon was in possession of and were considered "ready-to-use." For example, Ramstein alone, which hosts the 86th Airlift Wing, occupies 3,100 acres and has a base replacement value of $3.1 billion. Because of the more contested nature of the U.S. presence in Turkey, defense planners had been perhaps more reluctant to make similar investments in the infrastructure in Anatolia. Both the bases and ports in Turkey needed to be upgraded, particularly those in the southeast. Furthermore, the Pentagon wanted to open up a northern front, which would include the deploying of between 60,000 and 80,000 troops through Turkey. As outlined in previous chapters, social unrest in Germany had placed constraints on the ability of the United States to operate there, but had never led to bases being closed or troops being withdrawn as a result of protest. For this reason, in the run up to the Iraq War Pentagon planners had less cause for concern about obtaining basing access in Germany. Turkey, however, was considered less reliable and would have to be persuaded. In what follows, I show that obtaining basing access in Germany was not as easy as the Bush Administration had assumed and that it proved to be impossible in Turkey. I first discuss the case of the Bundesrepublik.

Obtaining Basing Access in Germany

Until now, most of the literature on U.S.-European relations during the Iraq War has focused on the conflict within NATO. A number of publications on U.S.-German relations during the Iraq War have focused on explaining why Berlin did *not* support the war.[16] Henry Kissinger referred to the dispute within NATO as the "gravest crisis in the Atlantic Alliance since its creation five decades ago."[17] More critical literature such as David Harvey's *The New Imperialism* has also pointed to the possible creation of a Berlin-Paris-Moscow axis in opposition to Washington, indicating the potential revival of

[16] Szabo 2004; Gordon and Shapiro 2004; Pond 2004.
[17] Daalder 2003.

inter-imperial rivalries.[18] I would agree with these assessments in that the episode in 2003 was the worst crisis in U.S.-German and U.S.-Turkish relations
since the end of the Cold War. However, as indicated, the German government
in fact *did* cooperate with U.S. war plans by providing basing access.

The Pentagon started trying to ensure basing access in the summer of 2002
when Germany was in the midst of a national election campaign. Having declared
"unconditional solidarity" with the United States less than one year earlier, Gerhard
Schröder had begun to clearly distance himself from the Bush Administration in
August, as administration officials began escalating the rhetoric regarding Iraq.
Vice President Dick Cheney's speech on August 26, 2002 to a veterans' association included a call for regime change, regardless of the results of the weapons
inspections. Schröder was infuriated by the speech, not just because of the shift in
U.S. policies but because he had learned about it from the media. Having stood by
the United States thus far and having survived a no-confidence vote, he was angry
that he was not even informed ahead of time, much less consulted.[19] In an interview in *The New York Times* shortly afterwards, Schröder expressed his exasperation at what appeared to be the new line: "How can you exert pressure on
someone by saying to them: Even if you accede to our demands, we will destroy
you?" A close confidante of Cheney was asked whether they had considered the
impact the speech might have in Germany and the election there. His reply was:
"Why should he care about the reaction in Germany?"[20]

Such comments are perhaps further proof that the administration saw
Germany as a "toolbox" which could be used when necessary and not as an
ally that should be consulted. This is even more remarkable when one considers the fact that the entire invasion hinged on the military infrastructure in
Germany, and yet administration officials apparently never seriously considered the possibility that the basing access might not be granted.

In a speech in Hannover in August 2002, Schröder declared that the
Bundesrepublik would be following a "German Way" and not supporting the
U.S. war in Iraq which he described as a military "adventure."[21] On September
4 in Berlin, he reiterated his position: "Under my leadership Germany will not
participate in an intervention in Iraq."[22] One day later, however, the German
media reported that Schröder was not excluding the possibility that the United
States could have access to the military infrastructure in Germany.[23] The

[18] Harvey 2005.
[19] Szabo 2004: 25.
[20] Cited in Szabo 2004: 28.
[21] The term "*Deutscher Weg*" or "German Way" referred also to socioeconomic policies, meaning
that Germany would not succumb to the neoliberalism that characterized the American system.
Although the term therefore may have a positive connotation for Germans interested in preserving their welfare state, for neighboring European states it may have carried the connotation of the
"*Deutscher Sonderweg*" or "German special path," which refers to the historical development of
Germany as distinct from that of other European countries since the nineteenth century.
[22] http://www.lr-online.de/nachrichten/Top-Themen-Irak-Konflikte-keyUSA-Deutschland.
[23] Schwarz, Patrik, "Schröder ringt um Lufthoheit," *Die Tageszeitung*, September 5, 2002.

Pentagon had already made clear that it did not consider German soldiers essential. The military bases in Germany, however, were absolutely necessary for any military operation.

This triggered a debate about the legal aspects of the U.S. bases in Germany. Gert Weisskirchen, a member of parliament (SPD) argued that the United States had the right to use the bases which could not be denied them. On the other hand, Klaus Schwarz, an analyst at the Stiftung Wissenschaft und Politik (SWP), was of the opinion that the German government could of course deny the United States the use of the bases and its airspace. According to him, "it was the normal practice: you have to be asked, and you have to agree to give permission."[24]

The Chancellor himself did not appear very interested in clarifying the issue, avoiding the question during press conferences. Press speakers at both the Foreign Office and at the Ministry of Defense were told not to give out any information about the legal aspects of the treaties; a veritable *"Auskunftssperre"* or *"suspension of information"* was declared. Otfried Nassauer of the Berlin Information Center for Transatlantic Security (BITS) speculated that this could be attributed to the fact that increased activity had already been observed at the Ramstein and Frankfurt bases.

Apparently assuming that Schröder's hesitation to commit to the provision of basing access was related to the campaign, President George W. Bush asked Foreign Minister Joschka Fischer after his speech at the UN on September 12, "When is your damn election over?"[25] After the ballots were counted on September 23, Schröder had won by 6,000 votes, the closest election in the history of the Bundesrepublik. Although U.S. administration officials were well aware of the fact that Schröder's victory was largely attributed to his anti-war position, they expected unimpeded access to the bases in Germany, over-flight rights, and for German soldiers to protect the bases. Later, German pilots would even be asked to man the NATO Airborne early warning and control system (AWACS), which were to be deployed to Turkey. Initially, Schröder hesitated to give in to U.S. demands.

Having waited for a decision regarding basing access for more than six weeks, President Bush decided it was time to make a phone call to the German chancellor. According to the German news media, which had obtained transcripts of the ten-minute telephone conversation, Bush, who had never congratulated Schröder on his reelection, instead reminded Schröder of his father's (Bush senior's) achievements during the process of unification more than twelve years earlier. He also emphasized that Chancellor Kohl had promised him that Americans would continue to enjoy basing access just as they had during the Cold War.[26]

[24] Ibid.
[25] Szabo 2004: 29.
[26] "Bush verkohlt Kanzler," *Die Tageszeitung*, February 13, 2003.

On November 8, 2002, the UN Security Council passed Resolution 1441 which gave Iraq a final opportunity to comply with its disarmament obligations. All fifteen members of the Security Council unanimously voted to pass the resolution, demanding that Iraq cooperate "immediately, unconditionally, and actively" with the UN Monitoring and Verification Commission and the International Atomic Energy Agency, or otherwise face "serious consequences."[27] On the same day, Russia, China, and France issued a joint statement that Resolution 1441 included no automaticity in the use of force, which the American and British representatives confirmed.

Later that month, during the NATO summit in Prague, Schröder issued a statement that in the event of war, the United States would have unlimited access to its basing infrastructure in Germany, even in the event that the invasion was not sanctioned by the UN. He also agreed to release Bundeswehr soldiers from their regular duties so that they could protect U.S. installations.[28] In early December, the Greens held a party congress (*Parteitag*) and passed a resolution to the effect that the government could only provide basing access and overflight rights if the war was sanctioned by a UN mandate. Foreign Minister Joschka Fischer called his party members' resolution "yesterday's debate" (*eine Debatte von gestern*) and insisted that everything depended on Resolution 1441. Whether or not another UN resolution was necessary to sanction the use of force, he claimed was unclear.[29]

Until now, only the hawks in the Bush Administration had supported such an interpretation of Resolution 1441. But claiming that Bush did not need any further UN approval for the invasion of Iraq did not in any way mean that he did not require approval from Germany for the provision of basing access. It did, however, make it possible for Schröder and Fischer to obfuscate and argue that the issue was out of their hands. When asked whether Germany was obliged to provide overflight rights even without a UN mandate, Schröder said "the federal government does not comment on legal opinions, it makes politics."[30] In a television interview on December 12, Schröder reasserted his intention to provide basing access, but was ambiguous about whether he had the ability to do otherwise, declaring: "This means we cannot place any restrictions on the freedom of movement of our friends and allies, and where we could, we will not."[31] Fischer found a more precise, but equally ambiguous formulation: "*Kriegsbeteiligung nein, Bündnissicherung ja*" ("no to participation in the war, yes to securing the alliance").[32]

---

[27] http://www.state.gov/p/io/rls/fs/2003/17926.htm.
[28] "Schröder weicht vom Anti-Kriegs Kurs ab," *Der Spiegel*, November 21, 2002; "Clear as Glass against the War; Germans still agree to aid U.S. and Israel," *The New York Times*, November 28, 2002.
[29] "Der schwierige Umgang mit den Überfliegern," *Der Spiegel*, December 10, 2002.
[30] "Der Kanzler wird das Schiff nicht verlassen," *Der Spiegel*, December 11, 2002.
[31] Cited in Pflüger 2003.
[32] "Kreative Mehrdeutigkeit," *Der Spiegel*, December 16, 2002.

On December 16, the Party of Democratic Socialism (PDS) brought charges against Chancellor Schröder that he had agreed to participate in a war of aggression. The formal eighteen-page long complaint was submitted by Wolfgang Gehrcke, the foreign policy adviser of the PDS and Evelyn Kenzler, the legal adviser to the party.[33] This led to a formal review of the legality of the issues by Attorney General Kay Nehm.[34]

By January 2003, plans were underway to stage large anti-war demonstrations in mid-February. How large they would be, of course, no one knew. Speculations were made that a movement that did not oppose but rather supported the government could not be very large, because it was assumed that having no opponent in one's own government would dampen mobilization efforts.[35] In fact, not only did the ruling coalition oppose the war, but also the Christian Democrats (CDU) who formed the opposition. The Greens, who were in coalition with Schröder's SPD, perhaps sensed that it would be politically opportune to reiterate their official anti-war position and differentiate themselves from the Christian Democrats, Social Democrats, and everyone else who was opposed to the war. In mid-January, Reinhard Bütikofer, the head of the Green Party, declared that the Greens should spearhead the peace movement.[36] Peter Strutynski, who had functioned as the chairman of the national *Friedensratschlag* (Peace Council) for a number of years, said in an interview that this was a laughable statement. The Greens, he said, were of course invited to attend the peace rallies, but to declare themselves as leading the movement was not accurate, to say the least.[37]

A few days later, the trade unions chimed in. During an anti-war conference organized by several trade unions in Hessen, Frank Spieth, the chairman of the DGB in Hessen and Thüringen, criticized the national leadership and argued that the unions should head the movement. As in the 1980s, the idea of five-minute symbolic work stoppages was discussed, which then took place in March (see discussion later in this chapter).[38]

On January 30, the *Berliner Zeitung* publicized a report by the *wissenschaftlichen Dienst* of the German Bundestag, which is similar to that of the Congressional Research Service. The German Parliamentary Research Service

---

[33] The full text of the accusation (Strafanzeige) can be found at: http://www.unikassel.de/fb5/frieden/themen/Voelkerrecht/anzeige.html.

[34] Rath, Christian, "Per Strafanzeige gegen den Krieg," *Die Tageszeitung*, December 24, 2002.

[35] "Seit an seit mit verdächtigen Gestalten; nicht gegen die Regierung, sondern mit Rot-Grün auf die Strasse zu ziehen, verunsichert manche Friedensbewegte," *Frankfurter Rundschau*, January 24, 2003.

[36] Schwarz, Patrik, "Zurück auf die Strasse; Die Grünen geben sich wieder als Bürgerinitiative," *Die Tageszeitung*, January 18, 2003.

[37] "Seit an seit mit verdächtigen Gestalten; nicht gegen die Regierung, sondern mit Rot-Grün auf die Strasse zu ziehen, verunsichert manche Friedensbewegte," *Frankfurter Rundschau*, January 24, 2003.

[38] "An der Gewerkschaftsspitze fehlen die klaren Signale," *Frankfurter Rundschau*, January 20, 2003.

had concluded its report almost six weeks earlier, on December 18, 2002. Because the report was intended for official use only it could not be publicized, but a summary of the report can be found online. The review came to two major conclusions: (1) the United States did indeed need to request permission to use its military bases and second, and (2) the German government was in no way obliged to grant permission because the NATO treaty does not justify preemptive wars.

The Anti-War Movement in Germany

On February 15, 2003, anti-war demonstrations took place around the world, in places as diverse as Amsterdam, Baghdad, Buenos Aires, Cairo, Cape Town, Karachi, London, Madrid, New York, Rome, Seoul, Tehran, Tokyo, and Washington, DC.[39] An estimated 10–15 million people took to the streets, making it perhaps the largest protest event on a single day in human history.[40] Although the organizers were predicting that "only" 80,000 people would show up for the demonstration in Berlin, approximately 500,000 people marched in the German capital, not including the number of demonstrators in other cities, making it one of the largest protest rallies in the history of the Federal Republic.[41] The demonstration was remarkable not only for its size, but also because it was attended by a number of federal ministers such as Bundesminister Renate Künast, Jürgen Trittin, and Heidemarie Wieczorek-Zeul, as well as Bundestagspräsident Wolfgang Thierse and Frank Bsirske, the head of the Verdi services trade union. Fifty groups belonged to the "Action Coalition" (*Aktionsbündnis*) by February 15.

Another important event took place on March 14, when the European Trade Union Confederation (ETUC) called for Europe-wide work stoppages in protest against the invasion. Both the German and Turkish trade union confederations (DGB and Turk-İş/ DİSK, respectively) are members of the ETUC. About seventy trade union organizations in thirty-eight countries participated in 10–15-minute work stoppages, making it the first simultaneous work stoppage in Europe in recent history. In Germany, the DGB had already demanded in May 2002 that Schröder withdraw his offer of "unconditional solidarity" with the United States so as to avoid being entangled in a war with Iraq.[42] In

---

[39] For an illustrated account of protests in forty cities around the world, see Koch and Sauermann 2003. See also Howard Zinn, "A Chorus against the War," *The Progressive*, March 2003; and Jonathan Schell, "The World's Other Superpower," *The Nation*, April 14, 2003.

[40] For a comparison of the February 15 protests in eight different countries including the United States, the UK, Germany, Switzerland, Italy, Spain, Belgium, and the Netherlands, see Rucht and Waalgrave, 2010. For a detailed analysis of the anti-war protests in the United States in the aftermath of 9/11, see Vasi 2006.

[41] "Millionen protestieren in aller Welt gegen den Krieg," *Frankfurter Allgemeine Zeitung*, February 16, 2003. See also Rucht and Walgrave 2008.

[42] "Krieg ist keine Lösung, auch nicht für den Irak! Beschluss des DGB-Bundeskongresses im Wortlaut," May 2002.

December 2002, during a meeting of the International Metalworkers Union (IMB), Klaus Zwickel, the President of the IMB and head of the IG Metall, made a statement warning against the escalation of "hate and violence." The fact that the statement was given in the United States was seen as an important political signal, but it was also interpreted as a response to pressure from mid-level trade unionists who had founded the *Gewerkschaftliches Netzwerk gegen den Krieg* (Trade Union Network Against the War.) Peter Strutynski of the Peace Council commented that Zwickel's political allegiance was still closer to that of the governing SPD than rank-and-file union members.[43] On March 14, between 11:50 a.m. and 12 noon, several hundred thousand employees stopped work. Large companies such as Opel, Audi, and Daimler Chrysler were affected by the temporary work stoppage. In addition to workers in the automobile industry, postal workers, transport workers, public school teachers, and many other employees from North-Rhine Westphalia, Baden-Württemberg, Berlin, Bavaria, Hessen, Sachsen-Anhalt, and Bremen all participated in the collective work stoppage. Employees in Spain, Portugal, Austria, Switzerland, Greece, France, Belgium and a number of other countries also took part.[44]

Although Green members of the governing coalition had participated in the February 15 demonstration, a month later when the invasion began, the Greens were still debating the legal terminology, such as what exactly could be considered "support" for a war of aggression and its implications for international law. According to the Green spokesman for legal issues (*rechtspolitischen Sprecher*) Volker Beck, only actions which directly supported the intervention could be considered "support." The left wing of the Green Party, represented by Hans-Christian Ströbele and others, agreed with Dieter Deiseroth, a judge at the Federal Administrative Court (*Bundesverwaltungsgericht*), who has published on the legal dimensions of the Status of Forces Agreement (SOFA). According to Deiseroth, even indirect support was included; for example, when military jets used German airspace for the intervention that did not fly directly to Iraq but via another country. Even more far-reaching in terms of its ramifications was the question of whether the war was defined as *"völkerrechtswidrig"* (a breach of international law) and therefore illegal as Ströbele demanded or as merely *"völkerrechtsfeindlich"* (adverse to international law) as Volker Beck argued.[45] If Berlin would declare the war to be a breach of international law, and therefore illegal, this would mean that it could not allow the United States to use the military bases.

Shortly after the bombing campaign began on March 20, 2003, the Green Party convened in a *Sondersitzung* (special meeting). With Ströbele and Beck

---

[43] "'Ich warne vor einer Eskalation von Hass und Gewalt,' IMB Präsident warnt USA vor Alleingang im Irak – Mit einem Kommentar," December 11, 2002, http://www.ag-friedensforschung.de/themen/Gewerkschaften/zwickel.html.

[44] "'Es ist zehn vor zwolf. Ein Krieg steht unmittelbar bevor' Gewerkschaftsproteste gegen den Krieg – Eine beeindruckende Bilanz der Aktionen," *Einblick*, March 31, 2003.

[45] "Das Völkerrecht im Schwebezustand," *Frankfurter Rundschau*, March 21, 2003.

TABLE 4.1. *Protests and Blockades of U.S. Military Facilities in Germany February–March 2003*

Place	Number of participants	Organizer
Grafenwöhr	150	
Spinelli Barracks in Mannheim	Varied	Independent Opponents of War (*Unabhängige KriegsgegnerInnen*)
Frankfurt airport	1,000	Resist
Ray Barracks in Friedberg	25	High school students
Spangdahlem Air Force Base	500	
European Command in Stuttgart Vaihingen	6,000	
Geilenkirchen – AWACS Base		Local church, SPD, Greens

representing the two different wings, the majority of the Greens decided that it was "not yet clear" if or to what extent international law was being violated.[46]

In addition to the large demonstration on February 15, a number of blockades were organized in front of U.S. military bases. The blockades varied in size, but for the most part consisted of a number of people standing in front of the entrance to an installation carrying anti-war signs. Several blockades took place in February before the invasion began, but most of them happened during the month of March. In Mannheim, the Spinelli barracks were blockaded every Saturday for a period of more than six months. Table 4.1 provides an overview of these blockades. Compared to the forms of civil disobedience that the peace movement had organized in the 1980s, they neither featured prominent participants as in Mutlangen, nor were they truly disruptive as the maneuver obstructions were. With the exception of the blockade in Mannheim, they were also of short duration as opposed to the long-running blockades in Mutlangen that continued for a period of several years. Finally, whereas the blockades in Mutlangen in 1983 began several months before the missile deployment was scheduled, the blockades in 2003 took place several months after the crucial decision to grant basing access had already been made.

Bases in Germany – War in Iraq

According to a number of observers, because many major installations are located in Germany and because Schröder placed no restrictions on the use of

---

[46] "Die Grünen und das Völkerrecht 'Zur Zeit nicht klar,'" *Frankfurter Allgemeine Zeitung*, March 21, 2003.

these facilities, Germany in fact did more to enable the Iraq War than many other countries who belonged to the "Coalition of the Willing" and officially supported the war.[47] In fact, Joschka Fischer himself indicated as much in an interview on March 5: "Hardly anyone does as much as we do in the alliance. What are those who approve of a military solution doing? Other than approving, they're not doing much."[48] All told, Schröder agreed to the provision of basing access, the deployment of AWACS surveillance aircraft to Turkey, and the maintenance of chemical and biological warfare detection vehicles in Kuwait. German soldiers flew AWACS planes in Turkey from February 26 to April 17, 2003 as part of "Operation Display Deterrence." The airplanes were there to protect Turkey from a potential attack by Saddam Hussein; the mission included four airplanes, thirty-eight soldiers, and nine civilians.[49] The BND (the German intelligence agency) cooperated with the CIA by helping to identify targets in Iraq. Frank-Walter Steinmeier, who was responsible for the secret services at the time, later became Foreign Minister and hence suffered no consequences for the collaboration. The UK used the port of Emden in northern Germany for the shipment of troops and materials to Iraq. Schröder also allowed 4,200 German soldiers to be released from their regular duties so that they could protect the approximately eighty large U.S. installations located in the Federal Republic. The U.S. forces who were freed up were then available for deployment to Iraq or Afghanistan. By increasing the German contingent of the NATO and EU forces in Bosnia and Afghanistan, more U.S. soldiers could be sent into combat.[50] Finally, German airspace was used on a regular basis: in addition to the planes that were starting and landing in Ramstein and Spangdahlem, B52 bombers flew from bases in the UK over German airspace en route to Iraq. Tobias Pflüger has therefore argued that Schröder and Fischer were employing a double strategy: publicly appearing to be against the war while in fact doing everything they could to support it, short of sending German soldiers to the front.

And yet, Rumsfeld didn't think it necessary to acknowledge the German contributions to the war. He instead pursued what some have called a divide and conquer strategy: distinguishing between "old" and "new" Europe, referring to the meeting between Germany, France, Belgium, and Luxemburg in

---

[47] Gordon and Shapiro 2004: 171.

[48] The original quote in German is: "Wir leisten im Bündnis so viel wie kaum jemand. Was leisten denn die Befürworter einer Militäraktion? Wenn man da nachfragt, ist außer dem Befürworten nicht allzu viel festzustellen." Interview mit Bundesaussenminister Fischer in *Stern*, March 5, 2003, http://archiv.bundesregierung.de/bpaexport/interview/26/470426/multi.htm. Last accessed July 18, 2008.

[49] "Über Krieg und Frieden entscheidet der Bundestag," *Frankfurter Allgemeine Zeitung*, May 8 2008; and "Wer über Krieg entscheidet," *Die Tageszeitung*, May 7, 2008.

[50] Pflüger, Tobias: "Es war gut, ihnen nicht geglaubt zu haben: Die deutsche Beteiligung am Irak-Krieg" in: ak – zeitung für linke debate und praxis, Nr 504, March 17, 2006, www.imi-online.de.

April 2003 about developing the EU military force as a "praline summit," and comparing Germany to Libya and Cuba are just a few examples of his pointedly undiplomatic diplomacy.[51] Karsten Voigt, who was the Coordinator for German-American Relations in the Foreign Office at the time, was convinced that the Bush Administration never realized that Schröder had made a conscious decision to support the war, despite his rhetoric to the contrary.[52]

Battles in the Courtroom

In addition to the public outcry including the mass demonstration on February 15 and the much smaller base blockades, battles around the use of military bases in Germany for the Iraq War increasingly took place in the courtroom. Having regained its full sovereignty, it is perhaps not surprising that challenges to the U.S. presence in Germany have now taken on a legal dimension. As already indicated, a number of legal experts argued that any support for a war without UN approval would violate the German constitution that prevented the participation in or preparation for a war of aggression.[53]

The ruling on June 21, 2005 by the German Federal Administrative Court (*Bundesverwaltungsgericht*, abbreviated: BVerwG) that the Iraq War and Germany's support for it was cause for "grave concerns in terms of international law" was in several ways unprecedented.[54] The BVerwG argued that, as a neutral state, German territory was inviolable and that any act of war was not to be tolerated. Hence, it was "forbidden to move troops or convoys of either munitions of war or supplies across" German territory or for military aircraft of the conflict parties to use its airspace.[55] The BVerwG further explained that Germany's obligations under international law took precedence over the NATO Treaty, the NATO Status of Forces Agreement, and the Presence of Foreign Forces Convention, in direct contradiction to claims made by the chancellor and his defense minister in 2003. The decision by the BVerwG regarding the legality of the Iraq War represented the consensus of international lawyers. What is remarkable about the decision, however, is that this statement has been made for the first time by a court of law, and in this case by the highest German court for administrative law matters.

A second groundbreaking decision was made on May 7, 2008, when Germany's highest court ruled that the AWACS deployment in 2003 was

---

[51] Gordon and Shapiro 2004.
[52] Interview with Carsten Voigt, Coordinator for German-American Relations, in the German Foreign Office in January 2008.
[53] Prof. Dr. Gregor Schirmer: "Deutschland ein Aufmarschgebiet der USA für den Krieg gegen den Irak? Eine völkerrechtliche Expertise": http://www.uni-kassel.de/fb5/frieden/regionen/Irak/schirmer.html.
[54] Schultz, Nikolaus, "Was the war on Iraq illegal? The German Federal Administrative Court's Ruling of 21st June 2005," *German Law Journal*, Vol. 7, No. 1, 2006.
[55] Schultz, Nikolaus 2006: 34.

unconstitutional.[56] At the time, Schröder had called it a "routine NATO maneuver," which did not require approval of the parliament. In reality, Schröder and Fischer most likely wanted to spare the members of their own parties the embarrassment of having to approve the participation in a war which they had officially opposed.[57] This decision has been interpreted as modestly increasing the power of the parliament vis-à-vis the executive, ensuring that the Bundeswehr is a *"Parlamentsarmee,"* an army of the parliament and not a *"Regierungsarmee,"* an army of the government.[58]

Thirdly, the attempt to prosecute former Secretary of Defense Rumsfeld by German attorney Wolfgang Kaleck in cooperation with the Center for Constitutional Rights received much attention and generated ill will, including suggestions by various observers that Washington should threaten to close down the remaining facilities in Germany.[59]

Finally, the emergence of the Linkspartei (Left Party) is significant because it has united the previously alienated left traditions in East and West Germany, including the PDS which had brought charges against Schröder for participating in a war of aggression. Although the PDS charges were struck down by the Attorney General, the more recent successful court battles may encourage further attempts to rein in U.S. military activity on German soil. The Left Party has been focused primarily on pushing back against neoliberal reforms, but it has also demanded the withdrawal of Bundeswehr troops from Afghanistan and increasing the power of the parliament to decide about the deployment of soldiers. Officially the Linkspartei continues to advocate for the dissolution of NATO, although Gregor Gysi is said to have reassured the U.S. Ambassador that this was merely to prevent the "dangerous" unilateral withdrawal of Germany from NATO.[60] Although several of the leading figures in the Linkspartei, such as Oskar Lafontaine, were active in the anti-nuclear movement of the 1980s, they have not made the U.S. presence in Germany one of their primary concerns.

TURKEY

The dramatic shift in U.S. foreign policy and increased U.S. military activity in Central Asia and the Middle East in the wake of 9/11, which sent out shock

[56] "Über Krieg und Frieden entscheidet der Bundestag," *Frankfurter Allgemeine Zeitung*, May 7, 2008.
[57] "Karlsruhe bremst Sicherheitsrat aus," *Frankfurter Rundschau*, May 8, 2008.
[58] The ruling was not only an embarrassment for the previous red-green coalition, but also for the ruling CDU which had plans to create a National Security Council, that would have even further undermined the role of the parliament.
[59] Rabkin, Jeremy, "Rumsfeld Accused; anyone home in the State Department?," *Weekly Standard*, December 4, 2006.
[60] "Forderung nach Nato-Auflösung: Gysi plauderte über linke Placebo-Politik," *Der Spiegel*, December 18, 2010, http://www.spiegel.de/politik/deutschland/forderung-nach-nato-aufloesung-gysi-plauderte-ueber-linke-placebo-politik-a-735428.html.

waves to many parts of the world, did not seem to ruffle many feathers in Ankara, at least initially. Some observers even suggested that Turkey's decade-long experience in counter-insurgency operations against the PKK would make it a particularly empathetic ally in the U.S.-led "War on Terror." Whereas the strategic culture of the Federal Republic has been described as a "culture of restraint," the Turkish high command has maintained a high level of asser-tiveness and self-confidence, and consider themselves to be the true guardians of the republic and the doctrine of Kemalism. But even the powerful Turkish armed forces have not been able to ensure stability for their American allies and the U.S. presence in Turkey. In short, Washington knew that it would have to convince Ankara to let it use its bases.

After the tragic events of September 11, there was an outpouring of public sympathy in Turkey as in Germany. All of the major parties in Turkey voiced support for the operation in Afghanistan, including the pro-Islamist Felicity Party and the newly formed AKP (Adalet ve Kalkınma Partisi). As William Hale put it, "the Taliban had no supporters in Turkey."[61] On October 10, 2001 the Turkish Grand National Assembly (TGNA) easily passed a resolution authorizing the dispatch of troops to support coalition forces in Afghanistan. Anti-war protests were limited to small events. For example, the Adana Anti-War Platform, consisting of small parties and organizations on the left, held a protest rally in Adana which demanded that the nearby İncirlik base be shut down. Speakers argued that in the event of war, İncirlik and the city of Adana would likely be the first targets, as İncirlik was the most potent symbol of U.S.-Turkish cooperation.[62] Over the course of the next year and a half, this critique gained resonance among the population as many came to see the American military presence not so much as a guarantor or provider of security, but rather as dangerously undermining it, as the United States sought to heave Turkey to the brink of war with Iraq.

In June 2002, Turkey took over the command of ISAF in Kabul for six months, for which Ankara received $228 million in direct U.S. aid.[63] By this time, the Ecevit government was on its last legs, having taken the blame for one of the worst economic crises in recent Turkish history. Despite the fact that the government in Ankara was unraveling, U.S. officials made it clear that Turkey did not really have a choice about whether or not to support the administra-tions' war plans. On July 31, Mark Parris, the former ambassador to Turkey, told the Senate Foreign Relations Committee session on Iraq that "Ankara will not have the luxury of sitting arms folded should Washington go after

---

[61] Hale 2007.

[62] "İncirlik Üssünün Kapatılması İstendi," http://www.bianet.org/bianet/kategori/siyaset/5297/incirlik-ussunun-kapatilmasi-istendi.

[63] Onis and Yilmaz 2005: 281, Hale 2007, and "Turkey Awaits Benefits of Aiding War on Terror," *The Wall Street Journal*, November 22, 2002.

Saddam.... Ankara, for her own interests, will need to take part in the planning and implementation of U.S. plans."[64]

In late August, the Under-Secretary of the Foreign Ministry, Uğur Ziyal, visited Washington where he received a red carpet welcome. Secretary of State Colin Powell, National Security Advisor Condoleeza Rice, Undersecretary of State Marc Grossman, and Deputy Defense Secretary Paul Wolfowitz all met with him in person. He also spoke with Vice-President Cheney via video-phone. According to a CRS report, such high-level meetings with an undersec-retary are "beyond the dictates of normal diplomatic protocol."[65] One reason for the break in protocol could be because this was the first time that the government in Ankara was officially informed of the full extent of U.S. war plans and Turkey's role in them. The plan included the deployment of 250 aircraft and 75,000–80,000 ground troops to Turkey for the invasion of Iraq, which would require the use of İncirlik plus eleven other bases. This wish list came as something of a shock, as Ankara had assumed that the United States would only require the use of the airfields, as was the case during the 1991 Gulf War.

The official request for cooperation was submitted in September by the Office of Defense Cooperation in Ankara to the Turkish Chief of Staff. The fact that this request was submitted from the U.S. military to the Turkish mil-itary was perhaps not an explicit attempt to circumvent the parliament, but it was an indication of the direct contacts and close cooperation between the two defense establishments. As has been outlined previously, one result of the social unrest during the Cold War was to partially democratize the issue of the foreign military presence in Turkey. In contrast to the benighted confusion in Germany, it was public knowledge that the parliament had to approve the deployment of foreign troops and not the Chief of Staff. This was codified in Article 92 of the Turkish constitution. Furthermore, it was also clear that the NATO alliance could not be used as a justification for supporting the United States, as the alliance was based on collective defense. In other words, there was no clause in the NATO treaty which could be used to coerce member states into participating in a war of aggression.

Because many of the sites that the United States wanted to use were either considered substandard or needed to be expanded for the large-scale opera-tion, it was deemed necessary to first inspect the sites and then upgrade them where necessary. In other words, the U.S. request for access to the bases on Turkish territory was divided into three steps: site inspection, site preparation, and the physical use of these sites for the military operation.

In a memorandum drawn up on October 15, the Foreign Ministry pointed out that such a large-scale operation would be unprecedented in the history of the Turkish Republic, and that it would be strongly opposed by public

[64] Cited in Kurkcu 2002.
[65] Migdalovitz 2002.

opinion.[66] At this point, it is important to keep in mind that the request for military cooperation was still being discussed behind closed doors and had not yet become public knowledge. The fact that a historical comparison was drawn is also significant. At its peak during the Cold War, the U.S. presence consisted of up to 30,000 soldiers, whereas the deployment for the Iraq War was calling for more than twice that number, albeit on a temporary basis. More importantly, during the Cold War, the U.S. military was not using Turkey as a launch pad for full-scale combat operations.

Communication between the American and Turkish militaries continued at the highest level. In late October, General Hilmi Özkök, the Turkish chief of staff, visited Washington, where he met with General Tom R. Franks, head of the Central Command and General Joseph W. Ralston, Supreme Allied Commander in Europe.[67] Undoubtedly, the issue of Iraq, and in particular northern Iraq, was high on the list of Özkök's priorities. Partially as a result of the no-fly zone over northern Iraq, the Kurdish population had gained a great deal of autonomy in Baghdad. Although in a de jure sense, a Kurdish state had neither been created nor recognized, de facto the Kurds had many of the trappings of a state apparatus, including their own armed forces. Northern Iraq was under the control of two militia groups: the Kurdish Democratic Party and the Patriotic Union of Kurdistan. It was an open secret that there were already between 2,000 and 5,000 Turkish troops in northern Iraq and that cross-border operations had been going on intermittently for a number of years.

Around the same time as Özkök's visit, the Congressional Research Service prepared a report entitled "Iraq: the Turkish Factor." The report outlined that Turkey had a number of political, economic, and humanitarian reasons to oppose the Iraq War and to therefore deny basing access. First, Turkey wished to preserve Iraq's territorial integrity and avoid the creation of a Kurdish state. Second, the government was afraid that a war would worsen the already precarious Turkish economy. Third, Ankara wanted to avoid an influx of Kurdish refugees and ensure that Iraqi Turkomen would be represented in any future government. The report was most likely intended to explain the necessity of gaining Turkish support through promises of financial aid (which Congress had to approve) and political backing in the EU accession process. However, the report assumed that despite these "misgivings," Turkey would in the end support the war and provide basing access because it valued the alliance with the United States and would want to help determine Iraq's future. Perhaps most tellingly, the report concluded that: "This will likely happen due to the influence of the Turkish military and foreign policy bureaucracy no matter which party or parties win the November 3 national election."[68]

Contrary to the memorandum by the Turkish Foreign Ministry, which clearly saw that the planned deployment for the Iraq War represented a break

[66] Hale 2007: 99.
[67] "Turkey Negotiates Role in War," *The Washington Post*, October 22, 2002.
[68] Migdalovitz 2002.

in the historic record in the sense that it would entail a temporary but mas-sive enlargement of the U.S. presence and an escalation of various direct and indirect threats to Turkish security, the CRS report presented the Iraq War as a continuation of previous forms of military cooperation.

From a politico-military standpoint, Turkey had until now cooperated with post–Cold War U.S. foreign policy, including U.S. operations during the First Gulf War and the twelve-year enforcement of the no-fly zone, and the Turkish military – still a strong actor in domestic politics – had never been shy about voicing its support for U.S. demands. From an economic standpoint, Turkey was still recovering from the financial crisis in February 2002, one of the worst in its history. For all of these reasons it would seem that the regime (regardless of who was elected – as the CRS report pointed out) had many reasons to cooperate with the U.S. plans for Iraq. Furthermore, the social forces from below that had crippled U.S. military activities in Turkey during the Cold War, were long forgotten and showed no signs of reemerging any-time soon. The Turkish left was no longer united in a single party, nor was it represented in parliament, as the Labor Party was during the 1960s. Instead, the left had become divided into microscopic sectarian groups that fought among themselves instead of seriously challenging the more established par-ties. The number of civilian employees who worked for the U.S. military had been drastically cut, making labor strikes on the military bases less likely and much less debilitating if they did happen. And the existing guerrilla organiza-tions, primarily remnants of the PKK, had a history of targeting Turkish and not American institutions. Furthermore, the PKK had declared a ceasefire in 1999. The generation of activists from the 1960s who had protested the U.S. presence had suffered various forms of repression: some had been banned from political participation, others had been imprisoned, and still others had fled the country, many of whom continued to live abroad in exile. This gen-eration, one could assume, had learned its lesson, whereas the younger gen-eration was not as politicized and did not grow up with the experience of a large-scale U.S. military presence. Finally, the topic of Iraq was not even a major election issue. Economic issues, instead, dominated the elections in November 2002. The AKP's platform did not make any serious commitment regarding Iraq, beyond wanting to preserve its territorial integrity. This was clearly in contrast to Schröder who took on an anti-war position and whose speeches were openly defiant of the U.S. administration. Furthermore, Turkish attention was focused on the EU. The European Council was due to meet on December 12–13 to decide whether to start accession talks. On August 2, the Turkish Grand National Assembly had passed legislation that abolished the death penalty and granted the Kurdish minority certain linguistic rights, such as the right to teach and broadcast in the Kurdish language.[69]

[69] The abolishing of capital punishment was also significant in terms of the Kurdish question. Among other things, it meant that the imprisoned leader of the PKK, Abdullah Ocalan, would not be executed, but that he would serve out a life-long prison sentence. Kurkcu 2002.

In sum, from the American point of view, it would seem that there was little that could go wrong. From the Turkish perspective, however, things looked different. Bülent Ecevit, veteran statesman and four-time prime minister of Turkey, handed over the premiership on November 19, less than three weeks after the CRS report was completed. His last advice to Abdullah Gül was not to get involved in Iraq.[70]

The election in November 2002 was remarkable in a number of ways. First and foremost, the secular Kemalist establishment was replaced with a new moderate Islamist elite in the form of the Justice and Development Party, the AKP. Furthermore, the AKP had won by such a landslide that it was able to form a government on its own. For the first time in twelve years, Turkey was not ruled by a coalition government, making the possibility of stability look promising. For the war planners, this meant that the AKP could – if it chose to do so – enforce party discipline through a group vote and thereby guarantee that at least their members of parliament would vote "yes" to the U.S. demands, ensuring parliamentary approval of the troop deployment. Finally, Recep Tayyip Erdoğan was in the peculiar situation of being the head of the AKP but not able to hold office. While he was mayor of Istanbul during the 1990s, he had cited a poem during one of his public speeches that allegedly idealized Islam. Because of a clause in the Turkish constitution, he was therefore unable to hold office. He was not officially able to become Prime Minister until mid-March 2003; in the interim period, Abdullah Gül would lead the government.

As soon as the newly elected officials were in place, the war-planners back in Washington lost no time in seeking to curry their favor. In mid-November, President Bush made a phone call to the EU President, Danish Prime Minister Anders Fogh Rasmussen in which he stressed the importance of Turkey's EU membership.[71] Cheney and Rice lobbied the Senate to approve Turkish trade zones, or "qualified industrial zones" as exist in Jordan and Israel.[72] In addition, the United States had been supporting Turkey's IMF reform program and had ensured that since 1997 Iraq sends half of its oil exports through a trans-Turkey pipeline.

As *The Washington Post* pointed out, the mere existence of talks between the United States and Turkey demonstrated the importance of Turkey to the Bush Administration. Of all the countries in the region which had agreed to provide basing access, only Turkey and Jordan were offered aid. Hence, the offer of economic assistance was already unusual. The fact that none other than the President, Vice-President, and National Security Adviser were personally

---

[70] Hale 2007: 99.
[71] "US Discusses Aid for Turkey to Defray Costs of an Iraq War," *The Washington Post*, November 19, 2002.
[72] "Turkey Awaits Benefits of Aiding War on Terror," *The Wall Street Journal*, November 22, 2002.

involved in securing this assistance or even lobbying on the part of another country, was even more unusual. The reason for this was directly related to the issue of basing access. In order to force Saddam Hussein to fight a war on two fronts, it was necessary to plan an invasion from the north and from the south. Basing access to the south of Iraq had already been secured in a number of the small Persian Gulf emirates (this is discussed later in this chapter). To open up a northern front, geography dictated that only three countries came into question: Iran, Syria, or Turkey. Ever since the U.S. military had lost its foothold in Iran with the revolution in 1979, U.S. officials had little hope of returning. Syria was similarly out of the question. In short, if there was to be a northern front at all, Turkey was the only option.

By early December, it appeared that basing access had been secured throughout the Middle East. *The New York Times* ran a story entitled "Iraq's Neighbors Seem to be Ready to Support a War" and discussed the allegedly favorable positions of the governments in Kuwait, Bahrain, Qatar, Saudi Arabia, Iran, Egypt, Jordan, and Turkey. Of these, only Iranian leaders were openly opposed to the war.[73] In particular, Qatar, Kuwait, Saudi Arabia, Oman, and Bahrain had already agreed to let their countries be used as bases for the invasion. However, Saudi Arabia was reluctant to allow combat operations to be flown from its soil. During the 1991 Gulf War, eleven out of the twenty-three bases that were used for launching combat troops and aircraft were located in Saudi Arabia. Without the Arabian bases, it was necessary to secure access elsewhere, increasing Turkey's significance. Furthermore, if fighter jets were limited to bases south of Iraq, they would require refueling to bomb targets in northern Iraq.[74]

With the exception of Saudi reservations, it would appear that the rest of the Arab states had fallen into line. It was now necessary to get Turkey on board. After Bush, Cheney, and Rice had already done their part to secure Turkish cooperation, it was now Wolfowitz's turn. In early December, the Deputy Defense Secretary paid a visit to Ankara with the war plans in his briefcase. En route, he made a statement that highlighted what was at stake:

Turkish participation, if it does come to the use of force, is very important in managing the consequences, in producing the result as decisively as possible, and also in helping to make sure that post-war Iraq is a positive force in the region, not a destabilizing one. So it's very crucial to have Turkey intimately involved in the planning process. It can make a big difference for every one, especially for Turkey. It's not simply some favor that we're asking the Turks to do for us. It's something that can make a much better situation for Turkey, should it become necessary to use force.[75]

[73] "Iraq's Neighbors Seem to be Ready to Support a War," *The New York Times*, December 2, 2002.
[74] "Key Aide Seeks Military Pledge from Turkey," *The Washington Post*, December 3, 2002.
[75] Statement by Deputy Secretary of Defense Paul Wolfowitz en route to Turkey on December 3, 2002. The full statement can be found at: http://www.defenselink.mil/transcripts/transcript.aspx?transcriptid=2822.

Clearly, the administration was trying to make the case that the invasion of Iraq was good for Turkey, at least in the long run. Shortly after his visit, the plans for the invasion including the three-step approval processes of site inspection, site preparation, and actual deployment, were published in the Turkish press on December 13 and became public knowledge for the first time. In an article entitled *"Zor Karar"* (Hard Decision) it was reported that the United States wanted to deploy 90,000 troops through Turkey and to use six air bases and two ports. Having only been in power for one month, the AKP would now have to make what some have described as one of the hardest decisions in the history of the republic. To avoid taking responsibility for such a momentous decision, Erdoğan had suggested that a referendum be held. The article also reported that the list of demands that Wolfowitz presented was in fact already modified since the request that was submitted earlier in the fall, as a result of the government's unease. Originally the United States had wanted to use four sea ports: Iskenderun and Mersin in the Mediterranean and Trabzon and Samsun in the Black Sea. Because of the geographical location of the ports in the Black Sea, the AKP was doubtful of their utility for the Iraq operation, but assumed that it was more about a show of power in the Caucasus. As a result, the United States had agreed that the ports in the south would be sufficient. In addition to the six air bases that would be used for air strikes (İncirlik, Muş, Batman, Diyarbakır, Çorlu, and Afyon), eight other airports would be used for logistics, making a total of fourteen air bases that the United States wanted permission to use.[76]

The Anti-Base Movement in Turkey

From a historical perspective, the anti-war protest on February 15, 2003 was remarkable not only for its international coordination, but because of the timing of the event. In contrast to the Vietnam era, when the highpoint of mobilization was reached several years after U.S. involvement had begun, the protest on February 15, 2003, took place more than one month before the invasion began. Activists in Turkey started organizing even earlier. The first major demonstration in Turkey was held on December 1, 2002 in Istanbul, more than three and a half months before the invasion, and even before the precise nature of the deployment plans had been publicized. The anti-war platform in Istanbul was organized under the name the Anti-Iraq War Coordination, and was the main organizing body of the major demonstrations before and during the war. The demonstration brought together more than 10,000 people from 160 political groups and parties and made concrete demands regarding the use of Turkey by the U.S. military: "We demand that Turkey not take part in the war and not open its air space and the İncirlik Base for the death weapons of U.S.A." This demonstration and this particular demand is important for three reasons.

---

[76] "Zor Karar," *Hürriyet*, December 13, 2002.

First, it was not just a general expression of pacifist sentiment, but contained an explicit appeal to refuse granting overflight rights to the United States for the war. In Germany, only a small fraction of the peace activists made a similar demand. Secondly, it is important to highlight that their demands focused on the issue of air space, as the planned deployment of 80,000 troops was still being kept secret. Finally, the demonstration would seem to have proven correct the memorandum of the Foreign Ministry from October 15, which had predicted that the public reaction would not be enthusiastic.

Another large rally was held in Ankara on December 23, 2002. Big rallies organized by anti-war coalitions in major cities were supplemented by press releases and smaller protests throughout Turkey. There was a demonstration in Konya, a traditionally conservative/Islamic city, which attracted 20,000 people and was the largest protest rally in the history of the city.[77] According to one count, there were more than twenty-six anti-war coalitions in twenty-six cities; each of these coalitions included numerous anti-war groups under their umbrella.[78]

A month after Wolfowitz's visit, the TGNA had still not given permission for a 150-person team to inspect the military sites. *The Washington Post* reported that after making initial expressions of support, Erdoğan was now increasingly making "confusing" statements. Murat Mercan, the AKP party's deputy chairman, explained that his party was deferring to public opinion.[79]

Splits within the AKP

On January 27, the same day that Hans Blix briefed the UN Security Council on weapons inspections, Ertuğrul Yalçınbayır, the Deputy Premier of the AKP, was apparently convinced that the United States would go to war against Iraq regardless of the outcome of the weapons inspectors, and said "Why are you going to make a war like this against someone who has surrendered?" echoing a sentiment shared by Chancellor Schröder.[80] Other AKP members went beyond verbal opposition and joined the activists. Fatma Bostan Ünsal, a senior female member of the AKP, planned to travel to Baghdad on February 8 to act as a human shield, an event organized by an American Gulf War veteran. On the other hand, Bülent Arınç, the speaker of the Parliament, was also opposed to the war, but said that the AKP should not be swayed by the street, an indication that some factions of the party wanted to become part of the establishment, and not part of the opposition.[81]

---

[77] Interview with Yıldız Önen on June 7, 2005 in Istanbul.
[78] Interview with Kaya Güvenç in Ankara.
[79] "After calls on Turkey, US put on hold; Heeding Public opposition, Ankara delays decision on use of bases against Iraq," *The Washington Post*, January 8, 2003.
[80] Caliskan and Taskin, January 2003.
[81] Ibid.

Additionally, the AKP made a show of denying British requests to use Turkish bases. In other words, the AKP was being both populist in its anti-war rhetoric, and yet trying to gain what it could from the United States, while at the same time also trying to peacefully solve the conflict through its own initiatives. Gül had invited the leaders of the Arab world to a summit in Istanbul to put pressure on Saddam Hussein to cooperate, which did not, however, result in any joint action. In short, the AKP was split. Many had speculated that the AKP would be under pressure to please the secular establishment and pointed out that in the past, Islamists have had to make compromises. In 1996, for example, Erbakan, whose Refah Party descended from the same line as the AKP, received instructions from the high command to sign a cooperation treaty with Israel, which he then did despite his own objections. Even his cooperation did not appease the military establishment, and he was then deposed by a soft coup on June 18, 1997. For all of these reasons, even those observers who may have been sympathetic to the anti-base/anti-war cause did not have much hope that the AKP would stand up to the United States.

Surrounded as the AKP is by the small, yet violently powerful ruling civilian-military bureaucracy, which has resigned itself to getting what it can from war, it is naïve to expect the party to resist the impending conflict actively.[82]

Despite the Turkish delays, by January 11, Rumsfeld had already signed three deployment orders to the Middle East, totaling 87,000 troops, including soldiers, sailors, airmen, and marines.[83] On January 17, *The Washington Post* reported that Turkey was close to allowing 15,000 troops to deploy through the southeast, a number that was much smaller than the 80,000 initially envisioned by the administration. According to the article, 80,000 soldiers was "too much for the Turkish public to digest" and there were "concerns that the U.S. troops would never leave." According to an opinion poll taken in March 2003, 83.7 percent of those polled believed that the United States would permanently station its forces in the region and would not leave after overthrowing the Ba'th regime.[84] When General Richard B. Myers, the chairman of the Joint Chiefs of Staff, flew to Ankara a few days later, he was greeted by several thousand anti-war demonstrators.[85] From the Turkish perspective, the threat was perceived as not merely stemming from the Kurds in northern Iraq or from Saddam Hussein's alleged arsenal of weapons of mass destruction, but also from the deployment of 80,000 U.S. soldiers through southeastern Turkey, which would turn Anatolia into an expansive military base and launch pad for the war. Although the United States was an official ally of Turkey, it was seen as

---

[82] Ibid.
[83] "US Force in Gulf is Said to be Rising to 150,000 Troops," *The New York Times*, January 12, 2003.
[84] Uslu, Toprak, Dalmis, and Aydin 2005, table 15.
[85] "In Brief," *The Washington Post*, January 20, 2003.

the aggressive party that was not only starting a war in Turkey's backyard, but also trying to drag the Turkish people into it as well. What many commentators and perhaps even administration officials did not realize was that Turkey was not only worried about economic consequences or Kurdish refugees, but that it felt threatened by the war. So strong was this threat perception that Ankara feared that Saddam Hussein may retaliate against Turkey and launch an attack. For this reason the government requested additional protection from its NATO allies in the form of AWACS planes that could detect and potentially intercept any missiles launched against Turkey.[86] This episode illustrates how even the preparation to execute concrete harm against one's adversaries can result in collateral harm for one's allies. It also illustrates how the U.S. presence had evolved into a prototypical protection racket, as defined by Tilly and discussed in the introduction. Although the U.S. may indeed have offered protection to its ally from the threat of retaliation, the threat itself had arisen largely as a result of U.S. actions. General Özkök summarized this general sentiment well when he said: "I find it hard to understand that those beyond the oceans, who say they are threatened, do not believe Turkey when it says it faces the same threat from right across its border."[87]

By early February, still more than six weeks before the invasion, members of the First Infantry Division based in Germany had already received orders to deploy to Turkey as soon as Ankara gave the green light.[88] On February 6, the TGNA voted on the issue of allowing the preparation of the bases for war. In a closed session, which was most likely intended to hide the identity of those voting, 308 voted in favor and 193 were opposed, with 9 abstentions. Before the vote Erdoğan had announced: "Our moral priority is peace, but our political priority is our dear Turkey."[89] The AKP had thus supported the resolution. Renovations then began on the Turkish bases of İncirlik, Diyarbakır, and Batman and at the airports of Çorlu, Sabiha Gökçen in Istanbul, and Afyon.[90]

Organized Labor and the Iraq War

Trade unions also used their leverage to organize anti-war events. Work stoppages in protest of the impending war took place on January 16 in Kocaeli, the province directly east of Istanbul, and in the city of Samsun, located along the Black Sea Coast.[91] On January 22, the administrative board of DİSK, the more

[86] "Irak savaşında Türkleri koruyacağız," *Hürriyet*, December 13, 2002.
[87] Taskin and Caliskan, March 27, 2003.
[88] "Turkey and US Closer on Troops; Council Gives Assent, with Condition," *The Washington Post*, February 1, 2003.
[89] Taskin and Caliskan, March 27, 2003.
[90] "Work Can Begin on Bases; Turkey, US Reach Terms for Renovation of Military Sites," *The Washington Post*, February 10, 2003.
[91] "Türkıyede Savaş Karşıtı Hareketler," www.bianet.org/bianet/yazdir/1618.

left-leaning of the two labor confederations, assembled in Diyarbakır for the opening of a local branch. At the meeting, the support of TÜSIAD, the employers' association, for Turkish involvement in the Iraq War was condemned.[92] Five days before the crucial vote, representatives of the major professional associations (TMMOB), as well as DİSK, and KESK, the public service employees union, and the Ankara Anti-War Platform met with parliamentarians to discuss the upcoming decision regarding the troop deployment, known in Turkish as *tezkere*.[93]

After the vote on March 1, anti-war labor activities continued, with a peace march on March 14, from Silopi and Gaziantep to Iskenderun, the harbor that was intended to be used as part of the deployment strategy. On March 21, one day after the war began, a two-hour work stoppage took place in Istanbul. In addition to DİSK and KESK, the more mainstream Turk-İş confederation also participated in the work stoppage.[94] On March 27, more than 2,000 people called in sick in Mersin, then marched to the city center chanting anti-war slogans, such as "The Iraqi people are not alone."[95]

Public Opinion

Between December 2002 and September 2003, two public surveying companies, Anar and Pollmark, were conducting opinion polls. According to a poll in January 2003, only 7.7 percent of the Turkish population was in favor of providing airspace and basing access for the intervention, whereas 77.8 percent thought Turkey should remain neutral.[96] When asked a month later, in February, an overwhelming majority still opposed the deployment of U.S. troops through Turkey (77.9 percent).[97] In poll after poll, the negative attitude of the Turkish population toward the war could be connected to the issue of the U.S. military presence. Three-fourths of the population (74.3 percent) were opposed to a U.S. military general having the highest authority in Iraq.[98] As mentioned previously, 83.7 percent thought that the United States would not leave Iraq after the war was over. When asked if they thought that Turkey should have given permission for the troop deployment to get financial support, 72.3 percent said "no."[99] When asked in March 2003 if the United States wants to deploy troops to Turkey again in the future, 75 percent said that Turkey should not accept this demand.[100]

---

[92] "DİSK Barış için Diyarbakırda," www.bianet.org/bianet/yazdir/16188.
[93] "Savaş Karşıtları Meclis Kapısında," www.bianet.org/bianet/yazdir/16892.
[94] "İşçiler ve Memurlar Savaşa Karşı İş Bıraktı," www.bianet.org/bianet/yazdir/17569.
[95] "Kamu Çalışanları Savaşa ve Bütçeye KarşI," www.bianet.org/bianet/yazdir/17736.
[96] Uslu, Toprak, Dalmis, and Aydin 2005, table 28.
[97] Ibid, table 35.
[98] Ibid, table 17.
[99] Ibid, table 42.
[100] Ibid, table 43.

According to a different survey by the Ankara Social Research Center, 87 percent of the population opposed any military intervention in Iraq, whereas 94 percent opposed the use of Turkish bases and troops to attack Iraq.[101] This would mean that more people were opposed to the use of the bases and Turkey's support for the war, than were opposed to the war in general. Clearly, the widespread sentiment in the population went beyond an abstract expression of pacifist sentiments, but was very much related to the hard fact of the U.S. military infrastructure in Turkey. This was then reflected in a movement that is therefore better understood as an anti-base movement than an anti-war movement. These public opinion polls were no secret, and yet, not many in Washington seemed to understand either the implications or the origins of these sentiments. The Bush Administration seemed to believe it could secure the cooperation of its allies by simply continuing to insist that Iraq was a threat to the United States, without understanding that it was the war against Iraq that threatened Turkey.

The preceding chapters have attempted to illustrate the contradictory nature of the U.S. military presence, as it produces both threat and protection. These chapters have underlined how the production of concrete harm during the Cold War inadvertently led to the production of collateral harm. During the Cold War, grievances emerged gradually as the U.S. presence was transformed from a legitimate protection regime to one that was pernicious in nature. With a war looming right across the border, the stakes were even higher. Based on the opinion polls, it would seem that the potential grievances caused by the war were evident to the majority of the Turkish population, not just those who lived in the direct vicinity of the military facilities. Furthermore, one could no longer expect that the Turkish people should accept the collateral harm caused by the bases as a lesser evil when compared to the threat of a common enemy – as Turkey and the United States no longer shared a common enemy. Instead, Turks were deeply unsettled by the actions of their ally, the United States.

The overwhelmingly intense opposition of the public to the Iraq War is even more remarkable when one considers that neither the government nor the media was inciting anti-war sentiment. Until now, the AKP had authorized the modernization of military facilities, it did not object to the arrival of American weapons and soldiers in Turkish ports for the planned deployment to Iraq, and it would go on to propose a resolution in parliament to allow the troop deployment. The media had also been taking a pro-war line, so much so that Gül speculated that some journalists must have received part of the $200 million that the U.S. government had allocated to win foreign hearts and minds. The media coverage of the demonstration on February 15 is a case in point.

The city government of Istanbul, where the main rally was planned for February 15, prohibited a demonstration from taking place, citing security concerns. Instead, the organizers hosted a "press conference," which was

---

[101] Cited in Caliskan and Taskin January 30, 2003.

attended by 10,000 people. The day was significant not only because it was an internationally coordinated protest event, but it had a particular significance in Turkey as the fourth anniversary of the capture of Abdullah Öcalan. In contrast to the mainstream German media's coverage of the anti-war demonstration on February 15, the Turkish press coverage was markedly negative. Zaman, a moderate Islamist newspaper, reported that "Öcalan's shadow hung over the anti-war demonstration" and featured a picture of someone smashing a shop window.[102] *Hürriyet*, a conservative mass circulation daily, wrote that the peace demonstration was provoked by followers of Öcalan.[103] *Milliyet*, a centrist paper, focused on the protests in other countries, and did not mention the demonstrations in Turkey at all.[104] Everyone who had given speeches at the "press conference" was arrested, but then later released.

By this time, the United States was working with a backup plan. The administration, however, was still assuming that Turkey would vote yes, and it had already sent warships to the ports of Iskenderun and Mersin on the Mediterranean coast. If the parliament voted yes, the troops and their equipment would be offloaded at the ports, and then deployed to six tent cities that would be established in the southeastern provinces of Gaziantep, Malatya, Diyarbakır, Mardin, Batman, and in the Silopi district of Sirnak.[105] If, on the other hand, the Turkish parliament voted "no," tanks and other armored vehicles would have to travel hundreds of miles across the Iraqi desert from starting points in Kuwait, and the resupply of U.S. troops with food and fuel would be more difficult.[106]

By this time, the negotiations with Turkey had attracted a lot of attention. The London-based pan Arab daily *al-Quds al-Arabi* wrote: "Turkey will enter the political history books as a state that has engaged in the fiercest forms of blackmail in order to obtain the highest possible price in return for making its bases available for the attack on Iraq."[107] It also reported that the United States was getting fed up with financial "blackmail," and started looking to alternative bases in Romania. The U.S. media had referred to the negotiations as a "Turkish bazaar" and Bush himself had said it reminded him of the horse trading of his native Texas.

---

[102] "Savaş karşıtı eyleme Öcalan gölgesi düştü," *Zaman*, February 16, 2003, http://arsiv.zaman.com.tr/2003/02/16/haberler/h21.htm.

[103] "Barış eylemine Apo provokasyonu," *Hürriyet*, February 16, 2003: http://webarsiv.hurriyet.com.tr/2003/02/16/249327.asp.

[104] Dünya, "Barış dedi," *Milliyet*, February 16, 2003: http://www.milliyet.com.tr/2003/02/16/index.html.

[105] "US Would Limit Action Kurds in Postwar Iraq; Vow to Turkey is Part of Deal on Troops," *The Washington Post*, February 27, 2003; and "In Turkey, Key Base Preps for US Troops," *Christian Science Monitor*, February 27, 2003.

[106] "If Turkey Finally Says No: A Hindrance, Aides Say, But There Are Other Plans," *The New York Times*, February 20, 2003.

[107] "Learning from Ankara," *Mideast Mirror*, February 24, 2003.

To be fair, the position of the AKP government was not unheard of in the annals of bargaining between states: they were simply trying to get some quid pro quo in return for their support, which would entail considerable economic and political costs. The anti-base movement, however, was operating against this logic. It was not demanding compensation for participation in the war effort, but was trying to pressure the government to refuse what was seen as a bribe, and to refuse the provision of basing access.

Whereas the February 15 demonstration was seen as the pinnacle of anti-war organizing in other countries, in Turkey it was merely one of many demonstrations, and perhaps not even the most important. One of the many other demonstrations was organized by the Human Rights Association of Turkey and was held in front of the İncirlik military base as the final stop of their "Peace Train" on January 24, 2003. The group, comprising approximately 100 human rights activists from around the country, made the following press statement: "Being aware of the fact that we cannot enjoy our rights and freedoms without peace, we oppose all wars on our own land or anywhere else in the world. We also oppose the use of İncirlik as a military base."[108]

A major international event was the Peace Weekend of January 25 and 26, 2003, organized by the Peace Initiative of Turkey (*Barış Girişimi*). Comprised mainly of prominent intellectuals, activists, academics, and journalists, the Peace Initiative had been formed soon after the September 11, 2001 attacks in New York and Washington DC, announcing that they did not "want to give in to either the power of terror or the terror of power" and that their main aim was to establish an unconditional movement for peace.[109] The Peace Weekend started with the Assembly of the 100s, which brought together 100 representatives from twenty professional groups including academics, writers, workers, artisans, engineers, lawyers, doctors, publishers and translators, cinema and theatre artists, business people, students, and the unemployed.

Two novel forms of activism were developed during this protest cycle.[110] One included a campaign to turn off the lights every day at 8 p.m. Although the actual deed of switching off the lights took place within individual households in private, the scene of entire buildings with simultaneously flickering lights became a very public display. Another form of activism which crossed the traditional public and private boundaries, included establishing direct contact with members of parliament by sending short messages over mobile phones. By mid-February, activists had gained access to all 550 of the personal mobile phone numbers of the parliamentarians. Kaya Güvenç estimated that on average, each parliamentarian would receive twenty to thirty messages a day, with increasing intensity towards March 1.[111] He remembers MPs telling him, "We

---

[108] "Savaş Karşıtları İncirlik'teydi," www.bianet.org, January 24, 2003.
[109] Mavioğlu and Günel 2005.
[110] For a more detailed description of the protests in 2003, see Altınay and Holmes 2009.
[111] Although Murat Yetkin's book hardly mentions the anti-war movement in general, in the discussion of the events of March 1, he makes a reference to the mobile phone messaging as a means of pressure that the parliamentarians were experiencing at the time. See Yetkin 2004: 170.

have had enough. We know what you want, so please stop." Finally, the simple fact that both secular and religious Turkish citizens were working together on a political campaign is perhaps a novelty in and of itself. The AKP did not officially mobilize its base to oppose the war, but it did not discourage individual people from participating in anti-war events either.

After the Turkish parliament had passed the resolutions for site inspection and site preparation, the Bush Administration believed it had been given a green light to proceed with its war plans. A team of more than 100 American logisticians, engineers, and communications experts had inspected the military sites, several million dollars had been invested in upgrading and preparing them for use, and more than 10,000 U.S. soldiers were waiting for deployment on battle ships off the coast of southern Turkey.[112] On the eve of the Iraq War, it would not be an exaggeration to say that İncirlik was viewed by defense planners as one of the prize jewels in the crown of the American empire of bases.

On March 1, 2003, the Turkish Grand National Assembly convened to pass the third and final step in the American war plan, which would open Turkish territory and airspace for the creation of a northern front against Iraq. It had finally been agreed that the deployment would allow a maximum of 62,000 soldiers, 255 warplanes, and 65 helicopters. As the members of parliament gathered, Marines in Iskenderun were already disembarking from their ships, while a crowd of 100,000 people had assembled outside the building in Ankara, demanding that the parliament refuse the U.S. request for basing access. As a result of a decision of the speaker of parliament, the final speeches of the parliamentarians were broadcast on television, increasing the accountability and therefore the pressure on the members of parliament. Out of the 550 members, 533 were present for the voting: 264 agreed to pass the resolution and 250 opposed it. Because the approval required a majority of those who were present, the 17 abstentions would count as "no" votes, bringing the total number of those opposed to 267. Approximately 100 of the AKP members had rebelled against their own party leadership and opposed the motion.[113]

In short, the Turkish Grand National Assembly had rejected U.S. demands, making it impossible to open a northern front for the invasion of Iraq. This outcome was not anticipated by the Bush Administration, the American media, Prime Minister Erdoğan, or even many members of the Turkish parliament, who remember the long moment of silence after the results of the vote were announced.

The Bush Administration did not recover quickly. The reaction on Capitol Hill has been described as a "gnashing of teeth" over the question of "Who lost

---

[112] "US Team to Visit Turkey in Sign of Shift by Ankara," *The Wall Street Journal*, December 12, 2002.

[113] Wendy Kristianasen, "Turkey: Post-Islamists in Power," *Le Monde diplomatique*, March 2003.

Turkey?"[114] Almost immediately, parallels were drawn to the crisis decade of the 1970s. *The Washington Post* called it a "stunning setback" and *The New York Times* described the situation as one of "extraordinary tension."[115] Even more interesting was the issue of how the outcome was explained. *The New York Times* blamed the outcome on the AKP, which it said was "too inexperienced to be able to count votes in Parliament or even within its own caucus."[116] Scholars writing with more hindsight have continued to explain the no vote as a result of the "inexperience" of the counter-elite. Philip Robins argued that the no vote was a result of "confused priorities, limited attention, and capacity overload."[117] Others have blamed militant Islam as being the culprit for the rise in anti-Americanism that seemed to burst out of nowhere.[118] But rather than inexperience or religion, it was the widespread mobilization against the impending war in Iraq that was the primary reason for the no vote. This mobilization did not take the character of a fundamentalist religious awakening, but rather that of an anti-war movement, similar to those in other Western countries. The movement in Turkey was, however, distinct from anti-war movements in Germany and elsewhere in that it had a concrete goal: to deny basing access to the United States.

As the soldiers withdrew from Turkish waters, the Pentagon shifted to Plan B which they knew would be "harder and uglier" without the northern front. Two years later, when the "harder and uglier" intervention turned into an all-out quagmire, Secretary of Defense Rumsfeld even blamed Turkey for the problems in Iraq:

> Given the level of the insurgency today, two years later, clearly, if we had been able to get the Fourth Infantry Division in from the north through Turkey, more of the Iraqi Saddam Hussein Baathist regime would have been captured or killed.[119]

In sum, Ankara refused to allow either its territory or airspace to be used as a corridor for the invasion, and the U.S. military presence in Turkey was rendered dysfunctional. However, the war planners had acquired basing access in a number of other countries within the region and the war was not prevented.

---

[114] Two years later, anti-Turkish sentiment in Washington was still making headlines, see "The Sick Man of Europe – Again: Islamism and Leftism Add up to Anti-American Madness in Turkey," *The Wall Street Journal*, February 16, 2005.

[115] "Turkey Rejects US Use of Bases; Vote Could Alter Troop Deployment," *The Washington Post*, March 2, 2003; "Powell Says US Can Wage War on Iraq without Turks," *The New York Times*, March 5, 2003.

[116] "Powell Says US Can Wage War on Iraq without Turks," *The New York Times*, March 5, 2003.

[117] Robins 2003.

[118] Taşpınar 2005.

[119] The remarks were made during an interview with *Fox News*, two days after the March 18, 2005 resignation of former U.S. Ambassador to Turkey, Eric Edelman, http://www.alternet.org/waroniraq/2005/03/003265.html.

David Cortright assesses the more general impact of the global anti-war movement:

> We have created the largest, most broadly based peace movement in history – a movement that has engaged millions of people here and around the globe.... The fact that this effort could not prevent war reflects not the weakness of our movement but the failures of American democracy and the entrenched power of US militarism.[120]

---

[120] Cited in Koch 2003: 116.

5

Conclusion

Losing Ground

I have argued that the U.S. overseas basing network represents a form of territorial acquisition unique among hegemonic powers. The small and fragmented basing network that was first assembled out of the "leftovers" of European empires in 1898 became truly global in scale after World War II. As the colonial era came to an end, the era of the baseworld was beginning. Whereas previous imperial powers deployed their soldiers and sailors to colonial possessions, the United States has been able to convince sovereign states to play "host" to U.S. troops and bases for what has become a longue durée, even during peacetime.

At times, the United States acquired overseas bases through conquest and occupation, as in Germany after the defeat of the Wehrmacht. Most installations within NATO, however, were created with the consent of the host country, as in the Turkish Republic. With the onset of the Cold War, the U.S. presence initially enjoyed widespread support in both Germany and Turkey, and indeed throughout most of Western Europe, as it was embedded in a multilateral framework and was seen as a guarantor that the conflicts that had engulfed Europe during the first half of the twentieth century would not be repeated. U.S. troops and bases safeguarded against both Soviet encroachment and the reemergence of German militarism, and were widely regarded as a public good at the onset of the Cold War.

Furthermore, precisely because the U.S. presence left the task of governing to the host country, was devoid of a civilizing mission or any pretentions to represent or be responsible for the welfare of the people it claimed to protect, it had rendered itself outside the democratic polity of its host countries. By renouncing responsibility for political representation and economic redistribution, perhaps the two single most common demands of social movements both in the global North and South, the U.S. overseas presence would seem to have had ingeniously insulated itself from the contentious politics that had defined the modern era. For all of these reasons, it was assumed that the U.S. baseworld would enjoy both legitimacy and stability. Despite these optimistic expectations, anti-base opposition emerged even in the face of a still-existing Soviet Union.

The purpose of this study has been to understand: (1) the causes of opposition to the U.S. military presence in Germany and Turkey from 1945 to 2005, (2) the various forms this opposition took, (3) the consequences or the impact on the U.S. presence, and (4) the larger practical and theoretical implications for the basing network as a whole.

WHAT WERE THE CAUSES OF ANTI-BASE UNREST?

Germany and Turkey are two key host countries or nodes in the global basing network. Both were front-line states during the Cold War, and the U.S. bases there, having existed for more than half a century, have become permanent fixtures in the respective geopolitical landscapes. Beyond these broad similarities, however, Turkey and Germany clearly represented very different host countries within NATO. Whereas Germany was defeated and occupied, Turkey's relationship with the United States began officially as one between allies, if not equals. In terms of culture, political regime, and levels of economic development, the two countries had little in common. At the end of World War II, Turkish society was characterized by a small Western-oriented elite and a large, mostly rural population. It was and still is a predominantly Sunni Muslim country that underwent periods of democratization interrupted by several military coups, whereas West Germany emerged from the period of occupation to become a wealthy, consolidated democracy with a mixed Catholic and Protestant population. And yet, social opposition to the U.S. presence displayed remarkable parallels in the two countries.

Given that we have found similar types of protest in very dissimilar cultural and historical settings, it would be erroneous to search for the cause of social unrest against the U.S. military presence as being the result of a distinct characteristic of the culture, religion, or political regime of the host country. Path-dependent explanations, such as those which argue that "liberating occupations" lead to stable basing environments,[1] cannot account for the emergence of the anti-nuclear movement, which at the time was the largest social movement in postwar German history and generated a serious crisis in bilateral relations. Other explanations for anti-base protests assume that criticism of the U.S. presence emerges primarily because of the politics of democratic transitions, but cannot adequately explain why anti-base movements may emerge in consolidated democracies.[2] Yet others see the level of security consensus among host

---

[1] Kent Calder has argued that liberating occupations lead to stable basing environments, with particular reference to Germany. See Calder 2007.

[2] Cooley 2008.

nation elites as decisive in determining outcomes of anti-base movements.[3] Finally, much of the previous scholarship has been focused on analyzing particular episodes of anti-base unrest, but lacks a longer-term perspective that allows one to explain how the U.S. presence has changed over time.

Instead of focusing on the nature of the political regime, or level of elite consensus within the host nation, my object of analysis has been twofold. I have first endeavored to analyze the American military presence itself: its policies, practices, and general modus operandi in each country. Second, I have studied the various forms of social unrest that emerged out of each society in opposition to the U.S. presence. In contrast to explanations that see the U.S. presence as either a static entity or somehow beyond scrutiny, I have developed a model that shows how the U.S. protection regime changed over time.

As discussed in Chapter 1, Charles Tilly distinguishes between legitimate protection (a ruler who provides a shield against a truly dangerous adversary) and a protection racket (a ruler who provides a shield against a threat that is either imaginary or that arose as a result of his own activities). For Tilly the primary independent variable was the nature of the outside threat, but in my conceptualization, there are two independent variables: the level of the outside threat and the level of collateral harm caused by the protector. For the U.S. presence to qualify as legitimate protection it must fulfill two minimum requirements: (1) it must offer protection against an outside threat which it did not create itself, and (2) it must be effective as a protector, including not causing harm toward those it claims to protect. If the U.S. presence deters the outside threat while causing harm to its protégés, it can be considered effective, but not credible protection. I refer to this type of protection regime as "pernicious protection." According to this definition, legitimacy is relational. It cannot be declared by the United States, but can only be bestowed on the United States by its protégés through active or passive support for the U.S. presence; however, the very same people can also call the legitimacy into question and withdraw their support.

During the Cold War, one of the two criteria necessary for qualifying as legitimate protection was automatically fulfilled through the existence of the Soviet threat. The second criteria, however, depended on the effectiveness and credibility of U.S. protection. During the early postwar period, both of these conditions were fulfilled and the U.S. protection regime could be described as "legitimate protection" and as hegemonic in the sense that it enjoyed either active or passive consent among its protégés in the host nation, or at least suffered no organized resistance.

In Turkey, this period of legitimate protection did not outlive the 1950s. The U.S. military began to lose credibility with the U2 spy plane incident in 1960, the Cuban missile crisis in 1962, and the Johnson letter in 1964. The behavior of military personnel as well as the normal operating procedures

---

[3] Yeo 2011.

of the U.S. Army, Navy, and Air Force also led to conflicts. The navy's shore leave policy, which involved anchoring in Istanbul or Izmir for the weekend and letting off approximately 4,000 U.S. sailors for rest and recreation, generated resentment and led to major protests against the Sixth Fleet. Turkish base workers, although paid better than their counterparts working in other sectors, complained of unfair and capricious labor practices. The fact that Americans occupied the best real estate in downtown Ankara, that U.S. military police patrolled the streets in these districts, and that the sprawling officers clubs, bowling alleys, and sport complexes were inaccessible to even Turkish base employees, were seen as problematic as well.

In Germany, the legitimacy of the U.S. protection regime lasted longer than in Turkey, but began to erode in the late 1970s, with the NATO Double Track decision. The deployment of nuclear missiles triggered a wide-spread sense of being "atomic hostages," and the enormous REFORGER exercises and war games which took place on private property, the low-altitude flights, the numerous plane crashes of military aircraft, and the constant noise from the firing ranges caused aggravation even among conservative voters.

Increasingly, the U.S. military was no longer seen as a benign or benevolent hegemon that protected against harm, but as *causing* harm, perhaps even more than it was allegedly preventing. Hence, the U.S. protection regime had changed from one that represented legitimate protection to one that can be considered a pernicious protection regime.

Furthermore, both the Turkish and German populations came to realize that enjoying U.S. protection meant that they had become de facto protectorates, and that existential decisions of war and peace were being made in Washington. Turkish and German citizens could neither vote in U.S. elections, nor could they vote American commanders in or out of office, although they were more or less permanently stationed in their country. The U.S. military presence was outside the realm of the host nation polity.

WHAT FORM DID THE OPPOSITION TAKE?

Opposition to the U.S. military presence led to several types of social unrest. The high point of social unrest in Turkey during the Cold War spans the period between the mid-1960s and the mid-1970s. Social unrest during this period can be divided into (1) protest within the parliament by the Türkiye İşçi Partisi, who declared their campaign against the U.S. presence as a second national liberation struggle after the War of Independence, (2) protest in the streets and civil disobedience as students demonstrated against visits of the Sixth Fleet, (3) strikes by base workers, and (4) armed struggle by the THKP-C and THKO, who kidnapped U.S. servicemen stationed in Turkey.

In West Germany, opposition to the U.S. presence during the Cold War can be divided into the same four broad categories: (1) protest within the parliament by Die Grünen who demanded an end to many types of military activities, with the withdrawal of West Germany from NATO being the final goal, (2) civil disobedience which took the form of base blockades and maneuver obstructions, (3) labor disputes among the unionized base workers and symbolic work stoppages in solidarity with the peace movement, and (4) armed struggle by the RAF and RZ which attacked American installations and carried out assassination attempts of high-ranking U.S. generals.

During the Cold War, the social movement sector in West Germany and Turkey displayed remarkable parallels, taking the form of four different categories of tactics: parliamentary opposition, labor unrest, violent attacks, and various forms of non-violent civil disobedience. In neither country was the opposition controlled by a single organization or ideological tendency, but in both cases became widespread, including large cross-sections of both societies. The heterogeneous anti-base groups that emerged employed a variety of tactics, so that virtually the whole spectrum of contentious politics was encompassed. These oppositional groups ranged from grass-roots citizens initiatives that were located next to small facilities in rural areas and were committed to non-violence, to internationally operating terrorist networks which carried out attacks against major installations. Clearly, these groups did not form a coherent whole in the sense that they represented a unified movement. On the contrary, activists often came from opposite ends of the political spectrum and saw one another's activities as detrimental to their own. Scholars have tended to study these groups separately. However, from the perspective of the U.S. Army, Navy, and Air Force, all of this represented a challenge to their authority and to the American military presence more generally.

Opposition to the U.S. military presence took the form of social unrest precisely because voice opportunities through official channels were generally lacking. Not only were voice opportunities lacking, but – short of emigration – exit options were as well. Finally, the possibility of appeasing the opposition through incorporation or cooptation was virtually impossible as well. Whereas activists whose demands are directed against their own government can potentially be incorporated into the existing political system, incorporation of anti-base groups into the realm of U.S. domestic politics was not possible. As non-citizens, the protectariat are excluded from the political system that claims to protect them. This does not, however, mean that they are incapacitated. On the contrary, I have argued that the protectariat possesses various potential sources of social power, including structural power, disruptive power, power of access denial, power of passive resistance, and power of armed resistance.

Finally, anti-base opposition should be understood as a distinct form of contentious politics, which may or may not overlap with anti-war movements. During the Cold War, anti-base opposition emerged largely independent of anti-war movements. The protest cycle in Turkey began before widespread mobilization against the Vietnam War, whereas in Germany the protest cycle began after the last U.S. troops were withdrawn and the civilian personnel were evacuated from Saigon. During the post–Cold War period, the most important anti-base protests in Turkey and Germany arose in the context of opposition to the invasion of Iraq. Some anti-base activists were motivated by internationalism, while others took issue with the American military presence (sometimes literally) in their own backyard. What this means is that there is no simple correlation between changes in protection regimes at the national level and changes at the systemic level in terms of U.S. power or prestige.

WHAT WERE THE CONSEQUENCES OF ANTI-BASE OPPOSITION?

During the Cold War, the external limits to both U.S. and Soviet military expansion were determined by the Yalta and Potsdam agreements, and were a matter of public record. The internal limits to U.S. military expansion and autonomy within its sphere of influence, however, were not as clearly or publicly demarcated. I have argued that these limits were determined by social forces within the individual host countries. In contrast to the role of Congress in questions regarding domestic bases, there is very little Congressional oversight of U.S. bases on foreign territory. If oversight has anything to do with seeing, then it is clearly those who live in the shadow of U.S. bases who see the activities that go on and have the ability to report on them, and thus exercise "oversight." Even those activities of the U.S. military abroad, which are not secretive, are often not seen by members of the U.S. Congress or the American media, but the local population does take note.[4]

At the most basic level, the result of anti-base opposition was to increase the amount of oversight and to decrease the autonomy of the U.S. military in each country. International relations scholars have referred to this general phenomenon as a "decoupling of basing ties," as discussed in the Introduction.[5] Although noticing the overall tendency, however, they have not explained how it has come about or why the "decoupling" happened to varying degrees in different places. As I have shown, the extent to which constraints were placed on the United States varied, despite the existence of very similar types of social unrest. What can explain the difference in outcome?

In Turkey, the U.S. protection regime broke down and became dysfunctional in the late 1960s and again from 1975 to 1978, as it became impossible for the U.S. military to continue to operate as it had in the past. In Germany, the U.S. military was obliged to change the way it carried out the annual maneuvers, to end low-altitude flights, and to implement certain environmental standards when new buildings were constructed. However, the protection regime never became entirely dysfunctional, as in Turkey. In what follows, I briefly review the four types of social unrest discussed in previous chapters and try to discern

---

[4] The CIA's secret rendition program, which began after 9/11, was discovered by airplane enthusiasts, whose hobby it was to watch planes taking off and landing. When they spotted an aircraft which they could not identify as belonging to any of the commercial airlines, they reported it.
[5] Harkavy 1989.

why similar types of protest resulted in different outcomes in terms of their impact on the U.S. presence.

Parliamentary Opposition

In both West Germany and Turkey, small parties on the left emerged which began to challenge the U.S. presence. Both the Workers' Party in Turkey and the Greens in West Germany played pivotal roles in acquiring and then pub- licizing information about the U.S. military presence that may have been kept secret otherwise. Both parties demanded that their respective countries with- draw from NATO.[6] Neither party was following a pro-Soviet line or criticizing the U.S. presence on the basis of an imported ideology, but both developed arguments and theories that were specific to their time and place.[7] Finally, both the Greens and TIP provided an institutionalized arena for challenging the U.S. presence. To be sure, these two parties were different for a number of reasons: they developed in two different eras and with different ideological persuasions, with TIP being an independent socialist party with close ties to the labor move- ment, whereas the Greens are an environmental party that developed out of the anti-nuclear movement. And yet, they played very similar roles in their respec- tive countries as they carried the banner of anti-base sentiment from the streets into the parliament.

TIP was established in 1961 and by 1963/1964 had defined itself as not only a party of and for the workers, but also as an anti-imperialist party. In fact, their definitions of socialism and anti-imperialism were closely intertwined, with Party leader Aybar even claiming that the primary contradiction was not between the working class and the capitalist class, but rather as being between U.S. imperialism with its collaborating landowners, compradors, and bureau- crats, and everyone else. In Turkey, TIP can be credited with having begun the campaign against the U.S. presence in the early 1960s, which was then taken up by others later in the decade. In Germany, the Green Party did not initiate the campaign against the U.S. military presence. The focus on security issues within the Green Party came only after the peace movement had already done the legwork of mobilizing around the issue of nuclear weapons.

For TIP, the struggle against the U.S. presence was framed as a second national liberation struggle that aimed to evict the "foreign occupiers" from Turkey. For the Greens, the struggle took the form of a pacifist and environ- mental reform movement that wanted to abolish nuclear weapons, to end the

---

[6] As discussed in Chapter 3, the Green Party dropped this demand from their program after German reunification.

[7] Bahro, one of the leading theoreticians of the Greens was an East German dissident and critical of the Soviet Union. Aybar was also critical of the USSR, to the extent that certain phrases such as "socialism with a human face" were reinterpreted in Turkish as "güleryuzlu Sozialism," or smiling socialism.

practice of low-altitude military flights, and to force the U.S. military to abide by certain environmental standards. In Germany, the call to evict the "foreign occupiers" was more nuanced and in many ways camouflaged. Whereas some such as Mechtersheimer demanded that all foreign militaries leave Germany, others demanded that West Germany withdraw from NATO or become reunified, both of which it was assumed would lead to the same result: the withdrawal of both Soviet and Western troops and bases from the two Germanies.

Although both parties were small, averaging between 3 and 10 percent of the vote, U.S. military officials were nonetheless worried, as they were a sign that the post-war pro-NATO consensus was eroding. As mentioned in Chapter 3, the Institute for Policy Analysis compared the Greens' statements that they would prevent the missile deployment to Nazi activities before the fall of Weimar.[8] U.S. Forces Liaison Officers (USFLOs) pointed to the creation of a new Ministry for Environment and Energy in 1984 and the election of Joschka Fischer to the state government of Hessen as institutionalizing the "tension" between the U.S. Forces and the Hessian authorities. Ironically, later observers would argue that these small parties contributed to the democratization of both German and Turkish societies.[9]

In Germany, the Greens were gradually transformed into a party which accepted militarism, both German and American varieties. In Turkey, the Workers' Party was abruptly silenced by the military coup in 1971. The misfortunes of TIP and the fortunes of the Greens have a lot to do with the different political opportunity structure of each country. However, the fact that the political opportunity structure of West Germany would appear to be more open than that of Turkey, does not necessarily mean that small parties there were more effective in the long-term. Della Porta and Rucht have suggested that the position of the center-left or social democratic party is important in deciding the fate of smaller parties – that is, whether they are coopted or repressed.[10] However, it was not the center-left party in Germany which adopted the militant anti-base demands of the Greens – although the SPD continued to oppose the missile deployment – but rather the center-left party in Turkey which, at least in part, realized the Workers' Party's demands to evict the U.S. presence by shutting down the bases between 1975 and 1978. Although the fate of neither party was determined by the United States, both results were amenable to U.S. interests. In neither Germany nor Turkey could an anti-base political party sustain itself for more than a decade.

How can we assess the impact of parliamentary opposition? Both parties were the parliamentary arm of social mobilization and were clearly effective in

---

[8] Although one cannot be sure, it is unlikely that the writers of this report feared the Greens as much as the Nazis were feared in the 1930s. It is more likely an attempt to discredit the Greens while it was still a young movement and potentially prevent it from becoming an established party.

[9] Hockenos 2008; Thomas 2003.

[10] Della Porta and Rucht 1995.

bringing the relevant issues from the streets to the chambers of the federal government. In some ways, however, they were more of an irritant for the larger mainstream parties than for the U.S. Forces, as they were the ones who had to answer questions and defend their pro-NATO politics. And yet often when inquiries were made, the large parties had to forward them to U.S. officials, as they sometimes lacked the relevant information and were not always able to provide answer themselves. USFLO memos have indicated that this was often a time-consuming activity.

In both Germany and Turkey, these opposition parties can also be credited with having nudged the larger social democratic parties to the left. In Germany, Chancellor Schmidt fell from power as a result of opposition to the NATO Double Track decision from within his own party, a decision he continued to support. Once in the opposition, the SPD joined the peace movement in opposing the missile deployment until they were removed under the INF Treaty in 1987.

In Turkey, the Workers' Party (TIP) had a lasting impact on the political spectrum even after it had been declared illegal. In addition to creating a political opening on the left, the social democrats (CHP) moved farther to the left in the 1970s under Prime Minister Ecevit. The decision by Ecevit in 1975 to suspend all activities at U.S. bases in Turkey is usually understood as an attempt to retaliate against the arms embargo declared by the U.S. Congress in response to the use of American-supplied weapons for the Cyprus intervention. He was also responding, however, to the domestic pressure created through the social unrest that began in the 1960s. The fact that the U.S. bases remained closed from 1975 to 1978 was a partial realization by a governing party of the declared goal of the opposition İşçi Partisi to eliminate the U.S. presence in Turkey.

By 2003, the left-wing opposition parties in both countries were either much weaker or much less oppositional than they had been during the Cold War. In the case of Turkey, the Labor Party had been declared illegal during the military coup in 1971. The transformation of the Green Party in Germany from a radical environmental and pacifist party to a party that has by and large adapted itself to mainstream Realpolitik has been discussed elsewhere.[11] The Green Party had already abandoned its pacifist tradition in 1999 when Joschka Fischer reinvented the lesson of World War II as no longer meaning "never again war" (*Nie wieder Krieg*) but instead "never again Auschwitz" (*Nie wieder Auschwitz*) to justify the intervention in the former Yugoslavia. Bundeswehr soldiers fought in combat for the first time since the end of World War II under the auspices of a Red-Green government. For this reason, large segments of the peace movement had a hard time believing that the anti-war position of the Greens in 2003 was genuine. The attempt to then position themselves as leading the anti-war movement was not taken seriously by anyone outside the party, with the head of the Peace Council calling this a "laughable" statement.

[11] Zirakzadeh 2006; Raschke 1993.

Instead, the Greens walked a fine line in publicly proclaiming their opposition to the war while purposefully avoiding the clarification of the issue of basing access and overflight. The report of the Parliamentary Research Service which found that the United States needed to request permission for basing access and that Germany was in no way obliged to grant this permission, provided the Greens with all the evidence they needed to make their case, if they had wanted to turn it into a political issue. The fact that there was a precedent for denying basing access, during the Yom Kippur War of 1973, was further evidence that the federal government – even as a semi-sovereign state during the Cold War – had all the power it needed to deny access. Instead of relentlessly uncovering secret information on military activities through the process of parliamentary inquiries as during the 1980s, the Greens were now purposefully trying to mislead the public and misinform them. Instead of using their power as members of the ruling coalition, they waffled between claiming to be against the war while in fact enabling the invasion.

In Turkey, the party which had initiated the campaign against the U.S. military presence in the early 1960s had been outlawed in the early 1970s. In the elections of November 2002, the moderate Islamist AKP party won a majority of the votes and could form the government without entering into a coalition with any other party. The Social Democratic CHP had obtained one of its worst results (19 percent) and formed the opposition. Hence, during the crucial period of the run-up to the Iraq War there was no left-wing opposition party in the parliament. The no vote in March 2003 hence testifies to the ability of social forces from below to influence politics without a parliamentary arm.

Armed Struggle

Instead of subsuming all forms of violent protest tactics under the catch-all phrase of "terrorism," it is useful for analytical purposes to divide them into four different sub-categories: (1) targeting of property (property damage), (2) targeting of combatants (guerrilla warfare), (3) targeting of specific categories of people (targeted assassinations), and (4) indiscriminate violence (terrorism).[12] If we are to understand how or why an armed organization can move from damaging property to killing indiscriminately, it is first necessary to differentiate between their activities.

In the early 1970s, the Red Army Faction and the Revolutionary Cells were focused on damaging or destroying property that belonged to the U.S. Forces. By 1979, the RAF had moved to targeted assassinations as they attempted to murder General Kroesen and General Haig. By 1985 at the latest, the RAF had become terrorists in the more narrow sense of the word as they had begun to engage in indiscriminate killings.

[12] For a discussion of these categories of terrorism, see Goodwin 2006.

The THKO and THKP-C followed a similar trajectory. The burning of Ambassador Komer's car in January 1969 was a first indication that protests had become violent. Dev-Genç was then formed in the fall of 1969, out of which the THKO and THKP-C were later formed. In November 1970, U.S. installations were attacked, and two Turkish policemen who were guarding the American embassy in Ankara were killed a month later, an indication that the groups had moved from committing property damage to targeted assassinations. In contrast to the RZ and RAF, the THKO and THKC carried out several kidnappings of U.S. or NATO personnel who were subsequently released unharmed. The kidnappings seemed to have been intended to demonstrate their own strength and hence intimidate the Turkish and American authorities and not to cause any physical harm to the individual G.I.s. In contrast, the RAF's killing of Edward Pimental to obtain his I.D. demonstrated their willingness to kill rank-and-file soldiers and alienated even many of their previous supporters.

The Turkish and German groups which were engaging in armed struggle thought of themselves – and were generally considered by others – as being the most radical or extremist left-wing organizations in either society. Some of the other activists saw them as belonging to the radical fringe of a common movement, and others saw them as having betrayed the goals of the movement and as belonging to an underground world which had little in common with their own. And yet, despite their radical ideologies and violent tactics, they did not have the most radical impact on the U.S. presence. Some U.S. commanders actually used the terrorist threat to urge the rank-and-file to stay the course and display their soldierly virtues. According to the "trip wire" strategy, U.S. soldiers were placed in harm's way so that if there were an invasion, American lives would be lost and the U.S. public could be rallied to participate in another war in Europe. American soldiers were constantly preparing for combat activity, but had planned for and trained for combat against the USSR, whereas they had comparatively little training in dealing with West German or Turkish militants. Precisely because they did not recognize the state's monopoly of violence and were engaging in violence themselves, the U.S. military no longer needed to treat them as citizens of an allied NATO partner who were deserving of protection.

What makes social unrest – and in particular armed struggle – in allied countries so intractable is precisely because it is happening in a territory defined as friendly and allied to the United States. If similar types of opposition were occurring in a combat zone, such distinctions may not necessarily be made. At this point, it is useful to recall the words of Major General Fox Conner, which were cited earlier: "Dealing with the enemy is a simple and straightforward matter when contrasted with securing close cooperation with an ally."[13] By becoming the "enemy," the RAF/RZ and the THKP-C/THKO had made

[13] Cited in Huston 1988: 128.

everything much more "simple and straightforward." The American, German, and Turkish authorities could easily agree that the armed militants – who they designated as terrorists – were a problem. Agreeing on what to do about the other types of social unrest was much more difficult. Because both the German and Turkish authorities saw the armed groups as a threat to their own authority, they did not need to be convinced to spend their own resources on counterterrorism. Not only did the violent attacks succeed in uniting the NATO allies, but it is not even clear that they would have been successful in achieving their goals had they remained divided on this issue. Whereas the Turkish groups perhaps carried out more kidnappings of Americans, the German terrorists were targeting the most powerful American generals in Europe. Had the assassination attempts on Generals Haig and Kroesen succeeded, things may have been different, but that would be speculation. It is clear, however, that the RAF was one of the longest-lasting terrorist groups in Europe, whereas the THKO and THKP-C had been fragmented and largely disbanded within a few years. In sum, the terrorist activities of the RAF and RZ were more prolonged, caused more property damage, and were more deadly than the activities of the THKO and THKP-C. If U.S. commanders had been intimidated by terrorist violence, they would have pulled out a majority of their troops from Germany, and not Turkey. Hence, the troop withdrawals in Turkey cannot be explained by the terrorist activity.

In one respect, the terrorist tactics did succeed in the sense that they caught the authorities off-guard. It is not that U.S. defense planners had never entertained the possibility of a terrorist threat, but had instead falsely defined it. According to the unit histories of the U.S. Army, U.S. officials had expected that terrorist groups would operate with the intention of seizing control of nuclear weapons. According to the available evidence, this was never the intention of either the Turkish or German armed organizations. Instead, their goal seemed to be to engage in a war of nerve, thereby intimidating the U.S. officers and soldiers, while hoping to incite other acts of violence among the general public. The fact that the tactics used were not expected (and therefore might be conceptualized as "novel") did not mean they were especially effective, as some social movement scholars anticipated.

American personnel became increasingly segregated from their host countries as a result of the violent attacks. Curfews were imposed, walls were built, security checks were increased, and it became increasingly difficult to have outside visitors on-base. Violence led to decreasing the amount of contacts between U.S. military personnel and the population of the host country while increasing the generalized sense of alienation and fear. The increased segregation made it difficult for the U.S. military's public relations machinery to work effectively, as locals may have had fewer negative contacts, but also of course had fewer positive contacts with soldiers and their families. Although some may have welcomed the "retreat to the barracks," others took it as evidence of

American indifference or aloofness toward their German or Turkish protégés, with whom many of them had earlier lived side-by-side.

By 2003, left-wing violence had largely become a thing of the past in both Germany and Turkey. In March 1998, the Red Army Faction declared that its twenty-eight-year-long "project" was finished and dissolved itself. Occasionally right-wing groups would carry out attacks against U.S. facilities, but there was no sustained violent campaign against the U.S. presence in the post-Cold War era as there had been during the 1970s and 1980s.

In Turkey, the THKP-C, and THKO had ceased to exist much earlier, with some splinter groups remaining in exile. The PKK, the major guerilla organization operating in Turkey, had been engaged in a prolonged civil war with the Turkish military, but had never posed a threat to the U.S. presence.[14] The United States did not even declare the PKK to be a terrorist organization until 1994. In 1999, after the capture of Abdullah Öcalan, the PKK declared a unilateral ceasefire, which it observed until 2004.

In both Turkey and Germany, despite the seemingly radical intentions, rhetoric, and tactics of armed militants, their actual impact on the U.S. military presence was less so. Compared to other protest tactics, violent attacks and kidnappings of U.S. military personnel proved to be relatively ineffective, both during and after the end of the Cold War.

Civil Disobedience

Civil disobedience can also be divided into sub-categories. In Germany, for example, some activities were relatively safe, whereas others were potentially dangerous for those involved. Some activities, such as the blockades in Mutlangen, were embraced by well-known personalities and entered into the mainstream repertoire of protest tactics, but other activities, such as the maneuver obstructions or breaking into military facilities, remained on the fringe. Some activists who engaged in civil disobedience thought of themselves as defending the German constitution, but others understood clearly that they were breaking the law and considered it necessary to do so. Some of those who engaged in civil disobedience enjoyed the protection of the police, whereas others did not.

During the protests against the Sixth Fleet in Turkey, the sailors were much more personally and directly affected than were the G.I.s in Germany by the base blockades. They were prevented from getting off their ships, had black paint thrown at them, and some of them were even physically assaulted, and generally made to understand that they were not welcome there. In Germany,

---

[14] The Center for Defense Information has a brief summary of various foreign terrorist organizations (FTOs) on its Web site. Under PKK, it lists "Main anti-US activities to date: none." http://www.cdi.org/terrorism/terrorist-groups.cfm.

the activists went out of their way to explain that they were not intending to harm or even criticize the U.S. military personnel themselves – even distributing pamphlets with slogans like "we like your face, not your base" and open letters. In both situations, the American soldiers were given explicit instructions to remain calm under all circumstances. The sailors were under orders not to fight back, and the army soldiers who encountered maneuver obstructers in Germany were told to leave the area as quietly as possible.

The U.S. military responded to civil disobedience in five ways:

1) Clearly, those who were engaging in non-violent civil disobedience could not be criminalized – and in fact could not even be openly criticized – by U.S. officials. Often the initial response of U.S. officials was simply to ignore or decline to comment on protest activities. By initially appearing to ignore the protests, the United States could create the impression that it was undeterred, and continuing steadfast in its mission. The port visits continued despite the protests, and the *Turkish Daily News* reported that it had been "proven" that the visits could take place. Only later were they then scaled back and discontinued altogether, so that it would not appear to be a response to the protests.

2) Because it was the job of the host nation's police force to protect U.S. installations and personnel during demonstrations, if a particular event got out of hand, it was possible to place the blame on the local police. In both countries, the U.S. officials who were observing the situations, tended to blame the German or Turkish authorities for not doing enough to protect the sailors or to prevent activists from breaking into military bases. This had the effect of making the police appear incompetent while deemphasizing the militancy or size of the protest event itself. Some of the liaison officers in Germany were perspicacious enough to point out that openly blaming the host country may be both counter-productive and undiplomatic.

3) A third response method was to coax local officials to make welcoming statements in public, which served to discredit the protesters and to create the impression of a harmony of interests between the two allies. In Turkey, the United States requested that Turkish authorities officially "welcome" the Sixth Fleet visits and sailors personally.

4) American officials chose to emphasize the NATO theme as a way of demonstrating commonality of interests. Sometimes this happened in subtle ways, as in Germany when the stationery used by the U.S. military was modified to include the NATO flag instead of the American and German flags. In general, the more the U.S. military was criticized, the more NATO was pushed to the foreground.

5) In some cases, the U.S. military did in fact make changes as a result of civil disobedience. Despite an extended period of recurring acts of civil

disobedience lasting over several years in Germany, the missiles remained in place from 1983 until 1987; that is, until they were removed as a result of the INF Treaty. In Turkey, the protests against the Sixth Fleet were also recurring but generally of shorter duration and less coordinated. And yet, these protests led to a cancellation of port visits. Clearly, more was at stake with the missile deployment, namely the balance of power between the United States and USSR, whereas the departure of U.S. sailors from the streets of Istanbul did not disrupt the global balance of forces. It did, however, mean that one aspect of the U.S. protection regime had been rendered dysfunctional, as shore leave was no longer possible in the Eastern Mediterranean. Finally, the symbolism of American sailors being tossed into the waters of the Bosphorus by unarmed protesters should not be underestimated. Perhaps more than any other, it was these images that came to symbolize the spirit of the student movement and the growth of the anti-imperialist left in Turkey.

Labor Unrest

Even more confounding than civil disobedience was the labor unrest of civilian base employees. I have argued that the overseas network of U.S. military bases forms a global division of violence and protection. Similar to production processes that occur through internationally connected commodity chains involving factories, warehouses, and retail stores, violence and protection are produced through the global deployment of troops and bases. The mere existence of soldiers and weaponry, however, does not automatically result in protection. Very few if any military bases are entirely self-sufficient, as most facilities rely on local labor to function. It is the seamless functioning and interoperability of military installations, air bases, sea ports, training grounds, and weapons depots that results in a protection regime that can credibly deter an outside threat and hence produce protection for one's allies while wielding violence against one's enemies.

As states came to rely on the mass of the population to fight in wars (rather than on paid mercenaries and professionals) they found it necessary to expand the rights of citizens to gain their cooperation in warfare. Giving workers the right to vote and the right to form labor unions were among the most important concessions that were granted to the lower and middle classes. In turn, these rights constrained the ability of states to engage in warfare. But when war-making capabilities are no longer located within a single nation-state but dispersed far and wide across the globe, there would appear to be no compelling reason to grant concessions to non-citizens. And yet as we have seen, an old problem is rearing its head once again: How to gain the cooperation of the population on which the military depends to operate? In other words, the problem posed by mass conscript warfare that was solved by the extension of citizenship rights is reappearing in another form as a result of mass anti-base protests.

In host countries, the United States may either own or lease the land on which military bases are constructed; it does not, however, own the local labor power. Although both taxes for land and rent for buildings can be waived by the government of the host country, the wages that are paid to local citizens cannot be waived, except in unusual circumstances as during the immediate occupation period in Germany when labor was supplied through requisition. Furthermore, the abolishing of the draft has made the United States even more dependent on local labor. Given the high degree of reliance on local labor, civilian employees have become the "critical cogs in the war machine" that worker-soldiers were as long as the draft existed. The structural power of base workers means that their protests are potentially much more explosive than any other form of social unrest.

As we have seen in the Turkish case, the strikes of base workers in 1967 and 1969 were key in leading to a breakdown of the protection regime. Whereas the kidnapping of individual military personnel by terrorist organizations was alarming, the loss of even a high-ranking general would not lead to the protection regime becoming dysfunctional. However, when it becomes impossible to land a plane at the major U.S. Air Force base in Turkey because of striking workers, when strikes are scheduled during NATO maneuvers, when they continue for up to six weeks and spread to numerous U.S. facilities across the country, causing family members to leave the country because the infrastructure they depend on is no longer operational, then it becomes clear that the military presence is no longer functioning. It should be pointed out that the strikes of base workers were not the most important strikes during this time in Turkey. The wildcat strike in Zonguldak in 1965, the general strike in Istanbul in 1970, and the May Day rally in 1977 were larger and more significant in terms of their impact on Turkish society. For the U.S. presence in Turkey, however, the strikes of the civilian employees had the most immediate effect, and contributed most to the protection regime becoming dysfunctional.

In Germany, the U.S. military was even more dependent on local labor than in Turkey, as it was one of the largest employers in the entire country, and the second or third largest employer in certain regions such as Rhineland-Palatinate and West Berlin. Potentially, labor unrest by German civilian employees could have been much more threatening than in Turkey. Despite the fact that the German unions were still strong during the 1980s, with the IG Metall winning a reduction in the work week from forty to thirty-five hours, and despite the fact that a number of individual unions supported the peace movement, the civilian employees organized in the ÖTV were more oriented to acquiring job security than engaging in political strikes against the American military presence.

Whereas labor unrest in the Turkish case was a big part of the reason that the U.S. military underwent a major transformation in Turkey, this was not the case in Germany. On the contrary, labor was a more stabilizing than a destabilizing factor in the German case. In the face of widespread opposition to the modernization of the U.S. Forces in Germany, U.S. officials argued that the construction of new facilities resulted in 6,000 new jobs during a period of high unemployment. Despite the relative acquiescence of German workers, the United States tried to replace local labor with dependents of U.S. personnel, in many cases the spouses of U.S. military personnel. This was known as "dependent hire" and allowed the U.S. Forces to be more flexible in the hiring and firing of employees who were much less likely to organize collectively against their employer. By hiring U.S. citizens, the U.S. military was less dependent on local labor and less vulnerable to labor unrest. However, both the government and labor organizations of host countries generally oppose this policy. From their perspective, the provision of jobs is one of the main benefits of a U.S. presence, especially if the bases are no longer seen as necessary in terms of providing security against an external threat. For a variety of reasons, it was never possible to fill all base-related jobs with dependents of military personnel. Hence, the reliance on local labor continues to pose a potential problem for U.S. authorities. From our comparison of the Turkish and German cases we can conclude that labor unrest by civilian base employees was the form of protest that had the most impact on the ability of bases to operate. The greater militancy of labor in Turkey explains why Turkey (but not Germany) moved from being a functional to a dysfunctional protection regime.

The collective European work stoppage on March 14, 2003 was a first. Although smaller in terms of participants than the demonstration on February 15, it was in some ways just as important, as the work stoppage threatened the economic interests of the employers. However, the work stoppage was more symbolic than actually disruptive, although the employers' associations opposed the industrial action.

In Turkey, organized labor contributed to the pressure on the government. In addition to their participation in the demonstrations, the union leaders also held a special meeting with their representatives in parliament just days before the vote. However, it was not until after the decision on March 1, and in particular after the war began on March 20, that the most important two-hour work stoppage took place. Finally, as the U.S. bases were not used for the invasion, there was no possibility that strikes by civilian employees could be used as a form of leverage. Therefore, although labor unrest during the Cold War was key in explaining the different outcomes in Turkey and Germany, it was not as decisive in 2003.

During the Cold War, the social movement sector in both Germany and Turkey displayed remarkable parallels. Protest against the U.S. military presence was

The Protection Regime and the Protectariat: Sources of Vulnerability and Sources of Power

Reliance on Host Nation for a functioning protection regime:	Type of protest	Type of social power
Labor (civilian employees)	Strikes, work stoppages, labor unrest	Structural Power
Civilian infrastructure for military activities (roads, harbors, airports)	Blockades	Disruptive Power
Civilian infrastructure for leisure activities (R&R, shore leave)	Civil disobedience (Sixth Fleet protests)	Disruptive Power
Territory	Refusal or restrictions imposed by parliament or civil disobedience (Manöverbehinderungen, making land off-limits for environmental reasons)	Power of Access Denial
Airspace	Refusal or restrictions imposed by parliament or civil disobedience (prohibiting low-altitude flights or stunt flying, restricting visibility)	Power of Access Denial
Access to military bases	Refusal or restrictions imposed by parliament or civil disobedience	Power of Access Denial
Civilian infrastructure for dependent support (off-base shopping, dependent schools)	Refusal to sell goods	Power of Passive Resistance
Safe and secure environment, law & order	Violent attacks; kidnapping	Power of Armed Resistance

highly differentiated between at least four forms of social unrest: parliamentary opposition, labor unrest, violent attacks, and various forms of civil disobedience ranging from symbolic blockades to high-risk activities. By 2003, many of these forms of social unrest had disappeared. Although the street demonstrations had reached record levels of mobilization, civil disobedience in Germany was limited to small symbolic activities, and violent attacks and labor unrest were insignificant to nonexistent. Parliamentary opposition was also less effective than in the 1980s, as the attempt to indict Schröder was struck down in the courts. In Turkey, the U.S. presence had been reduced to a skeleton of its previous size, making labor unrest by civilian employees an unlikely form of

leverage, as it was during the Cold War. Pressure from left-wing parties or guerrilla organizations did not factor into the political equation either. However,
protest against the Iraq War took more innovative forms than in Germany,
including the text messaging campaign that overwhelmed members of parliament to the extent that it went beyond mere petitioning and could be considered a form of civil disobedience. The cross-country caravan from Istanbul to
the İncirlik Air Force Base in Adana which traveled a distance of more than
1,000 km and the demands to turn the base into a children's park or women's
shelter were both more spectacular and more radical than the short blockades
in front of the gates of U.S. military facilities in Germany.

In terms of sheer numbers, the demonstration in Germany on February 15,
2003 was larger than any of the demonstrations during the height of mobilization during the early 1980s, with 500,000 people marching in Berlin that
day compared to 300,000 in 1981. This mass movement, however, was not
able to prevent the granting of basing access. Whereas the social unrest during the 1980s was able to put limits on what the U.S. military could do within
Germany, it was not able to achieve this in 2003. The paradox is that the movement in 2003 was not only larger, but also had more powerful allies in the government than in the 1980s, and yet it achieved less. This is not to diminish the
significance of the demonstration on February 15, 2003, which was perhaps
the largest protest event on a single day in human history, and by any standards a stunning achievement. But it is important to place this global protest
movement within different national contexts. In Germany, the protest did not
result in pressure on the government to deny basing access, but rather demonstrated support for the official anti-war position of the SPD-Greens. Hence, the
large number of SPD and Green politicians who were publicly on display at
the February 15 rally. During the large protests in 1981, pressure was clearly
being put on the ruling SPD to refuse the deployment of missiles. In Turkey,
the demonstration on February 15 occurred two weeks before the vote in the
parliament and therefore served as an important form of public pressure on the
government instead of boosting its moral authority as in Germany.

The question remains, however, why the movement in Germany did not
raise the issue of basing access to the same extent as in Turkey. In both countries, the U.S. presence no longer provided protection as during the Cold War.
Leaders in both countries as well as the population in general were critical of
the war because it lacked international legitimacy. The key difference could
be the level of collateral harm that the U.S. presence was believed to be causing in 2003, compared with the upsurge of unrest during the Cold War. The
deployment of Pershing II and cruise missiles was viewed as increasing the
likelihood of nuclear war; and the vigor of the 1980s movement stemmed
from the desire for self-protection. The war in Iraq, however, was not directly
threatening to Germany as it was to Turkey. Rather than being the central
front or battlefield during the Cold War, Germany was now merely a transit
point, en route to the Middle East. The thrust of the 2003 movement was

therefore based not on self-protection, but on the desire to protect others. In Turkey, however, these two movements for protection merged into a powerful force of opposition, to both the war in Iraq, and to the baseworld that makes all wars possible.

As I hope the historical evidence assembled here has shown, the U.S. presence overseas – even in friendly allied countries – has not been without its share of problems. A number of the dilemmas that arose in Turkey and Germany are endemic to the larger network of bases as a whole.

Controlling the Uncontrollable

Perhaps at the most fundamental level, the creation of an overseas network of military bases is an attempt to control what is essentially uncontrollable. By creating bases in foreign countries, the United States is at the same time attempting to create a monopoly on the use of force at the international level, and yet also relinquishing control of that monopoly to those states that host the bases. This means that the United States may make a unilateral decision regarding overseas bases, but that it cannot act unilaterally because it depends on overseas facilities such as Ramstein and Incirlik for the decisions to be executed. Not only does a functional protection regime require access to facilities, but also the territory, airspace, and often civilian infrastructure of host countries. It may depend heavily on civilian base workers or local businesses to provide goods and services. In short, a U.S. overseas military presence requires the cooperation of the protectariat, or the consent of the protected. And this is precisely what the United States cannot control. Possessing an immense network of overseas bases and being able to command the use of it are two different matters.

This is in some ways reminiscent of the situation of Great Britain during the inter-war years, which brings us back to our original comparison with the British basing system. In the words of Eric Hobsbawm: "Never had a larger area of the globe been under the formal or informal control of Britain than between the two world wars, but never before had the rulers of Britain felt less confident about maintaining their old imperial supremacy."[15]

Downsizing Does Not Solve the Problem

In some cases, the United States has responded to social unrest by attempting to assume a lower profile, which is essentially a euphemism for segregating U.S. servicemen and servicewomen from the host country, or by actually decreasing the number of troops or installations. Neither of these strategies has been particularly successful. In Turkey, the sprawling officers clubs, bowling alleys, and golf courses were no longer seen as the human dimension of the military presence, but, because they were off-limits to Turkish citizens, became objects that generated more resentment than goodwill. The attempts to camouflage the

[15] Hobsbawm 1994: 211.

military presence by allowing Americans to wear civilian clothing instead of their uniforms, or by painting the school buses for dependent children a different color, did not fool many people. Kent Calder's Contact Hypothesis would seem to be accurate: "We have found, in the world of cold, everyday life, that civil-military contact indeed predictably generates grassroots pressures that threaten to enflame base relations."[16] Defense planners thought that by making such contacts less frequent, they would decrease the likelihood of "enflaming" civil-military relations. However, pulling out of urban areas and resettling in more rural locations also did not solve the root cause of the problem.

By extension, smaller facilities or smaller numbers of troops does not necessarily result in smaller problems. The tiny installation in the village of Mutlangen became a lightning rod for social movements, whereas the gargantuan Ramstein base – the largest U.S. base overseas – was the object of fewer protests. Simply put, the issue is not the size or location of the U.S. presence, but whether the local population believes it is providing, or on the contrary, undermining their security.

Liaisons Can Only Do So Much

Wary of the fact that a large military presence may cause an occasional problem, a system of U.S. Forces Liaison Officers was established in West Germany after the end of World War II. These officers, who operated at the level of individual states (*Länder*) were mediators between the U.S. military and its German hosts. Whenever problems arose, they were the first to be sent forth with the task of sorting things out. Many liaison officers were stationed in Germany for ten years or more, while the political advisers in the State Department often had three-year assignments. Hence, they acquired valuable experiences and were able to gain the trust of their German counterparts as they had a relatively long-term perspective. The USFLOs can be credited with having solved a number of problems, and one can speculate whether the U.S. presence in Turkey may have been stabilized through the existence of such officers. Needless to say, the liaison officers were not omnipotent; they could not control the media, infiltrate activist organizations, or disperse protest events. Once the protest cycle in Germany was underway, the course of events was no longer under the control of a single organization, much less could it be manipulated by a single liaison officer. Ironically, those liaison officers who best understood what was happening in Germany were accused of "going native" and were subject to an outside investigation in the 1980s.[17] Apparently their inability to stem the tide of social unrest made some in Washington think they were not doing their job properly. By implementing the liaison system in a single country and by distrusting the political advice of the liaison officers, the United States was

---

[16] Calder 2007: 225.
[17] Interview with Michael Hanpeter, April 2, 2008.

limiting its ability to understand political change in those countries that host U.S. bases.

When Your Allies No Longer Need You

During the early period of the Cold War, both West Germany and Turkey could be characterized as security dependent, as they were incapable of defending themselves without American assistance. Because of their relative weakness, and because they both shared a long border with the Soviet Union, the threat for them was arguably greater than for the United States. Germany and Turkey needed the United States more than the United States needed them. But by the end of the Cold War, the relationship had been reversed. Now that there is no longer an outside threat looming across the Iron Curtain, Germany and Turkey no longer require U.S. protection. The United States, however, still has an interest in maintaining bases there. Furthermore, once the outside threat is gone, the legitimacy of the protection regime depends entirely on the United States and its actions. Mishandlings, excesses, or abuses of power can no longer be justified as necessary because of an outside threat.

When You Need the Consent of Foreign Citizens More Than Your Own

The expansion of the warfare state went hand in hand with the expansion of the welfare state for most of the twentieth century. Beginning with the crisis in Vietnam and continuing apace under Reagan, a decoupling of the warfare and welfare states could be observed. Because of popular resistance during the Vietnam War, the U.S. military began "delinking" from the U.S. population. The abolishment of the draft and the development of new technologies heralded the shift from labor-intensive to capital-intensive warfare. By abolishing the draft and moving to a professional military, the United States has to a large extent emancipated itself from its dependence on the consent and participation of its own citizens to succeed in war. The growth of private military contractors have taken this already anti-democratic tendency to a new dimension altogether.[18] However, I have argued here that the United States has not emancipated itself from its dependence on foreign territory, and hence on the consent of foreign citizens. In terms of its ability to wage war, the United States is in some ways ironically more dependent on the consent of foreign citizens than on its own citizenry.

DEMOCRATIZING AN UNDEMOCRATIC PROTECTION REGIME

Anti-base movements are not about the distribution of wealth or incorporation into an existing political system, but at their core they are about sovereignty – hence, they are generally not characterized by class polarization, but often by cross-class mobilization. Precisely because the main issue in the anti-base

---

[18] Singer 2003 and Scahill 2007.

movements is not wealth or the granting of civil liberties, it is not as easy to placate their demands. When workers organize collectively, it is usually possible to deflect potential militancy and incorporate workers through higher wages and to periodically "reincorporate" them through a continual increase in wages if necessary. In this sense, economic conflicts may be easier to manage because they can be continually drawn out with incremental increases in wages or benefits. Because anti-base movements are about reclaiming sovereignty from a foreign military presence, anti-base movements can be considered analogous to anti-colonial movements. If we stretch our temporal analysis, we can see that the first phase of empire, in which both land and labor belonged to the metropole, was undermined by powerful abolitionist or anti-slavery movements. The second imperial formation, characterized by ownership of land but not of labor, was subverted by anti-colonial movements. The anti-colonial empire of bases in which both land and labor are free, is now being eroded by anti-base movements in various corners of the globe.

By systematically analyzing both the causes and consequences of anti-base unrest, I hope to have made a contribution to understanding the limits of the U.S. basing network. These limits were set not by economic constraints or Congressional oversight or any other dynamics originating in the United States, but rather by social unrest in the countries that host U.S. military bases. Rather than a theory of expansion, I have offered a study of the empirics of contraction. Just as the decolonization of colonial empires took different forms, so too does the process of losing overseas bases vary over time and space. Having erected the largest overseas basing network in recent history, the United States is now losing ground.

LIST OF INTERVIEWS CONDUCTED IN THE UNITED STATES

1. James R. Blaker, former Deputy Assistant Secretary of Defense for Policy Analysis, Washington, DC, May 19, 2008.
2. Steven Cook, Fellow at the Council on Foreign Relations, New York, May 30, 2007.
3. Carine Germond, Professor of History at Yale University, written responses to interview questions, April 9, 2007.
4. Joseph Gerson, Director of AFSC Programs in New England, Boston, January 2009.
5. Michael Hanpeter, former U.S. liaison officer in Germany, Washington, DC, April 2, 2008.
6. Michael O'Hanlon, Fellow at the Brookings Institution, phone interview.
7. Chalmers Johnson, Professor Emeritus of UC San Diego, San Diego, August 2004.
8. Fred Ikle, Under Secretary of Defense for Policy under Reagan, phone interview, May 23, 2007.

9. Lawrence Korb, Senior Fellow at the Center for American Progress, Washington, DC.
10. Lt. Col. Karen Kwiatkowski, served at the National Security Agency and the Pentagon's Near East and South Asia Directorate, Mount Jackson Virginia, August 14, 2005.
11. Catherine Lutz, Professor of Anthropology at Brown University, Providence, January 2009.
12. Andrew Marshall, Head of the Office of Net Assessment in the Pentagon, Washington, DC, May 7, 2007.
13. Benton Moeller, former Head of the Liaison Officers in Germany.
14. John Mueller, Professor of Political Science at the Ohio State University, Washington, DC, May 17, 2007.
15. General William Nash, Fellow at the Council on Foreign Relations, Washington, DC, February 16, 2005.
16. Steven Szabo, Executive Director of the Transatlantic Academy of the German Marshall Fund.
17. Bryan Van Sweringen, former U.S. liaison officer in Germany, Washington, DC, March 20, 2008.
18. David Vine, Professor of Anthropology at American University, New York, August 2008.
19. Patricia Walker, Executive Director of the Overseas Basing Commission, phone interview, May 13, 2008.
20. Emira Woods, Foreign Policy in Focus, Washington DC.

ARCHIVE DOCUMENTS FOR GERMANY

This list does not include those archival documents that have been included in the Foreign Relations of the United States (FRUS) volumes that are published by the State Department. The FRUS volumes are included instead in the bibliography. Many of the documents listed have recently been declassified, hence, their original classification (whether secret, confidential, or top-secret, etc.) will not be indicated.

1. "Fulda-Gap: Hier könnte der dritte Weltkrieg beginnen," Friedensbüro Osthessen, 15. September 1984, Archiv Grünes Gedächtnis of the Heinrich Böll Stiftung in Berlin.
2. "Kommando über Europa: Das US EUCOM in Stuttgart-Vaihingen," Studienkreis EUCOM Stuttgart (editor), Stuttgart (undated, most likely 1989–1991), archive of the Hamburger Institut für Sozialforschung.
3. "Mutlanger Erfahrungen: Erinnerungen und Perspektiven," Wolfgang Schlupp (editor), Mutlangen, January 1994, private collection.
4. "Mordbase Ramstein," Die Bezugsgruppe Peaceniks (editor), Eigendruck, Saarbrücken, April 1984, Archiv-Aktiv in Hamburg.
5. "Der Fluch mit dem Tiefflug – oder: Widerstand ist moglich," in "Tiefflug: Die eigene Bedrohung," Starnberger Impulse, Mediatur Beilage, November 1988, private collection.

6. "Militärland Rheinland-Pfalz: Gesamtprofil der US-Streitkräfte in Rheinland-Pfalz," Die Grünen, May 1989, Die Grünen im Landtag Rheinland-Pfalz, Archiv Grünes Gedächtnis of the Heinrich Böll Stiftung in Berlin.

7. "NATO Herbst – Fulda Gap – Hildesheim – Manöver 1984," published by Antimilitarismus Kommission, Göttingen 1984, Archiv Grünes Gedächtnis of the Heinrich Böll Stiftung in Berlin.

8. "Generäle für den Frieden: 10 Fragen und Antworten zum Wettrüsten, zur 'Nachrüstung' und den Genfer Verhandlungen," edited by General a.D. Gert Bastian (BRD), General a.D. Michael Harbottle (GB), General a.D. Michael Meyenfeldt (Niederlande), undated pamphlet probably 1983, TtE Bücherei in Cologne.

9. "AirLand-Battle: Die neue offensive Strategie der USA," edited by International Physicians for the Prevention of Nuclear War (IIPNW), July 1985, TtE Bücherei in Cologne.

10. "AirLand Batte: Beängstigender Wandel der NATO-Strategie," Sozialistische Jugend Deutschlands, Die Falken, Hannover, private collection.

11. "Die Geschichte der NATO," edited by Bundeskongress autonomer Friedensinitiativen BAF, private collection.

12. "Bundesrepublik und NATO im Spannungsfeld: Eine Sammlung wichtiger Dokumente gegen die Kriegsgefahr," edited by Studiengruppe Bundesrepublik und NATO im Spannungsfeld, Berlin/Bonn, April 1984, archive of the Berliner Informationszentrum für Transatlantische Sicherheit in Berlin.

13. "Pershing II, Atomwaffen, Giftgas, Kriegsvölkerrecht: Untersuchung der Antworten der Bundesregierung auf zwölf Grosse Anfragen der Grünen im Bundestag," Bonn Juli 1984, archive of the Berliner Informationszentrum für Transatlantische Sicherheit in Berlin.

14. "In Gefahr und Höchster Not bringt der Mittelweg den Tod: Krise, Krieg, Friedensbewegung" Diskussionspapier der Revolutionäre Zellen and Rote Zora, archive of the Hamburger Institut für Sozialforschung.

15. Memorandum for: Commander in Chief, U.S. Army Europe and Seventh Army from the U.S. Forces Liaison Office, Rheinland-Pfalz and Saarland, October 11, 1988, private collection.

16. Memorandum for: Commander in Chief, U.S. Army Europe and Seventh Army from the U.S. Forces Liaison Office, Rheinland-Pfalz and Saarland, November 7, 1988, private collection.

17. Memorandum for Commander in Chief, U.S. Army Europe and Seventh Army from U.S. Forces Liaison office, Rheinland-Pfalz and Saarland on July 11, 1988, private collection.

18. Memorandum for: Commander in Chief, U.S. Army Europe and Seventh Army, from the U.S. Forces Liaison Office, Hessen on October 24, 1988, private collection.

19. Memorandum for Commander in Chief, U.S. Army Europe and Seventh Army from the U.S. Forces Liaison Office, Hessen "End-of-Tour Summary 1982–1988," private collection.
20. "Historical Review 1 January 1982 – 31 December 1983," Headquarters U.S. Army, Europe and Seventh Army, prepared by the USAREUR Military History Office, Forward by Bruce H. Siemon, Chief Military History Office, Commander's Statement by General Glenn K. Otis, May 1, 1985, archive of the U.S. Army Center for Military History.
21. "Strengthening NATO: Stationing of the 2d Armored Division (Forward) in Northern Germany" published by the Headquarters, United States Army, Europe and Seventh Army, May 22, 1980.
22. "To be or NATO be: Die NATO-Broschüre der Grünen," published by Die Grünen im Bundestag, Bonn, July 1988.
23. "Tiefflug: Die eigene Bedrohung," Starnberger Impulse, Mediatur Beilage, November 1988.
24. "Special Activities Report on Impact of German Industrial, Residential and Recreational Development on U.S. Forces in the Federal Republic of Germany," Memorandum for the Commander in Chief, U.S. Army Europe and Seventh Army from the U.S. Forces Liaison Office, Bavaria, March 31, 1981, private collection.
25. "Historical Review 1 January – 31 December 1981," Headquarters U.S. Army, Europe and Seventh Army, prepared by the USAREUR Military History Office, Forward by Bruce H. Siemon, Chief Military History Office, archive of the U.S. Army Center for Military History.
26. "Unit History: United States Military Liaison Mission in Potsdam to the Commander in Chief Group of Soviet Forces in Germany," 1984.
27. "Financial and Employment Impact of the US Forces on the German Economy in Fiscal Year 1983," report authenticated by Colonel Fredrick C. Schleusing, DCSHNA on June 5, 1984, private collection.
28. "The German Context within which USAREUR must live and operate" U.S. Army Europe and Seventh Army, Heidelberg 1983, private collection.
29. Letter from the Secretary of Defense Caspar Weinberger to Minister of Defense Manfred Wörner on October 16, 1987, private collection.

LIST OF INTERVIEWS CONDUCTED IN GERMANY

1. Karl Bredthauer, editor of Blätter fuer deutsche und internationale Politik, Bonn, July 11, 2007.
2. Andreas Buro, leading figure in the peace movement, Grävenwiesbach, May 28, 2006.
3. Dieter Deiseroth, written responses to interview questions.
4. Thomas Ebermann, founding member of the Green Party, Hamburg, 2006.

5. Joe Garvey, Public Affairs Officer for the U.S. Army Europe, Heidelberg, May 30, 2006.
6. Wolfgang Hertle, archivist at the Hamburg Institut für Sozialforschung and editor of Graswurzelrevolution, Hamburg, January 2007.
7. Klaus Layes, Mayor of Ramstein, Ramstein, July 3, 2003.
8. Wolfgang Kaleck, President of the Republican Bar Association and prosecutor in the case against Rumsfeld, Berlin, June 19, 2007.
9. Kalle Kress, peace movement activists and member of BUND, Kaiserslautern, July 2, 2003.
10. Lou Marin, peace movement activist, Cologne, January 2008.
11. Dr. Andrew Morris, historian with the U.S. Army Europe, Heidelberg, May 30, 2006.
12. Otfried Nassauer, Berlin Information Center for Transatlantic Security (BITS), Berlin, January 2008.
13. Norman Paech, foreign policy speaker of the Linkspartei, Hamburg, January 2007.
14. Tobias Pflüger, Member of Parliament for the Linkspartei, Brussels, January 2008.
15. Helmuth Priess, Chairman of the Darmstaedter Signal, Weilerswist, July 7, 2007.
16. Eva Quistorp, founding member of the Green Party, Berlin, summer 2006.
17. Karl-Heinz Roth, scholar and activist, Bremen, July 6, 2006.
18. Jürgen Rose, Lieutenant-Colonel of the Bundeswehr, Munich, January 15, 2008.
19. Dieter Rucht, WZB, Berlin.
20. Wolfgang Schlupp-Hauck in Mutlangen in 2006, one of the "permanent blockaders" in Mutlangen.
21. Erich Schmidt-Eenboom, activist, Munich, January 2008.
22. Hans Günther Schneider, farmer whose land was appropriated for the Spangdahlem Air Base, founder of BIEGAS citizens' initiative, Binsfeld, summer 2006.
23. Eckart Spoo, editor of "Die Amerikaner in der Bundesrepublik," Berlin, 2006.
24. Franz Steinkühler, former Chairman of IG Metall, written responses to interview questions.
25. Wolfgang Sternstein, activist and member of the Plowshares group, Stuttgart, July 2006.
26. Roland Vogt, founding member of the Green Party, Bad Dürkheim, January 2008.
27. Thomas Warth, trade union secretary for ver.di, Kaiserslautern, July 4, 2003.
28. Erhard Weber, Kaiserslautern, July 3, 2003.

29. Marianne Zepp, scholar and former activist, Böll Foundation, Berlin, January 2008.
30. Interviews with various U.S. military personnel stationed in Ramstein and other bases in Germany.

ARCHIVE DOCUMENTS FOR TURKEY

This list does not include those archival documents that have been included in the Foreign Relations of the United States (FRUS) volumes that are published by the State Department. The FRUS volumes are included instead in the bibliography. Many of the documents listed have recently been declassified, hence, their original classification (whether secret, confidential, or top-secret, for example) will not be indicated.

1. "Türkiye İşçi Partisi Büyük Kongresi" undated article published in Eylem, from the Kemal Sülker Collection, Box 604, of the International Institute of Social History in Amsterdam.
2. "Türkiye İşçi Partisi ikinci Büyük Kongresi" and "Basin Bülteni," November 20, 1966, from the Kemal Sülker Collection of the International Institute of Social History in Amsterdam.
3. "Genelge: Bütün İl ve İlçe Yönetim Kurulu Başkanliklarina," letter from Cemal Hakki Selek on October 14, 1966, from the Kemal Sülker Collection of the International Institute of Social History in Amsterdam.
4. "Current Status of Leftist Movements in Turkey" from Embassy to Department of State, July 11, 1970, National Archives and Records Administration.
5. Sixth Fleet Weekly Dispatch Summaries, U.S. Naval Archives
6. Memo from U.S. Consulate in Istanbul to the Secretary of State in Washington, DC, June 24, 1967, National Archives and Records Administration.
7. Memo from the U.S. Consulate in Istanbul to the U.S. Embassy in Ankara, October 7, 1969, National Archives and Records Administration.
8. Memo from the U.S. Consulate in Istanbul to the U.S. Embassy in Ankara, April 10, 1968, National Archives and Records Administration.
9. Memo from the U.S. Embassy in Ankara to the Secretary of State in Washington, DC, May 1, 1968, National Archives and Records Administration.
10. Joint memo from the State Department and the Department of Defense to Ankara and Istanbul, April 15, 1968, National Archives and Records Administration.
11. "Visits of the Fleet" published in Outlook on September 4, 1968, microfilm reel, archives of the Air Force Historical Research Agency.
12. Memo from the U.S. Consulate in Istanbul to the State Department, February 5, 1971, National Archives and Records Administration.

13. Letter from Major Frank E. Ruggles, May 13, 1969, National Archives and Records Administration.
14. Joint message from the Department of State and Defense Department to the U.S. Embassy in Ankara, May 23, 1969, National Archives and Records Administration.
15. Letter from Senator J. W. Fulbright to the Secretary of State, William P. Rogers, May 29, 1969, National Archives and Records Administration.
16. Report from the U.S. Embassy in Ankara to the Secretary of State in Washington, DC, June 18, 1969, National Archives and Records Administration.
17. Turkish U.S. Logistics Group (TUSLOG) report to the Headquarters of the U.S. Air Force, microfilm reel 3777, Air Force Historical Research Agency.
18. Letter from Secretary of State to the Embassy in Ankara, May 1969, microfilm reel 3777, Air Force Historical Research Agency.
19. Letter from the Chief of Staff of the U.S. Air Force to the Headquarters of the U.S. Air Force in Europe, June 1969, microfilm reel 3777, Air Force Historical Research Agency.
20. History of TUSLOG Det 170, July 1, 1968 – December 21, 1968, microfilm reel 3037, Air Force Historical Research Agency.
21. Memo prepared for the Under Secretary by the NEA from Joseph J. Sisco for a meeting with Turkish Ambassador Esenbel on April 29, 1970, National Archives and Records Administration.
22. Memo from Frank Cash, Country Director, Turkish Affairs, to the State Department in Washington, DC, January 16, 1971, National Archives and Records Administration.
23. Letter from H. G. Torbert, Jr., Acting Assistant Secretary for Congressional Relations to Senator John Stennis, August 25, 1969, National Archives and Records Administration.
24. "Three Hostages and Terrorist Gang Killed" Memo sent from U.S. Embassy in Ankara to the Secretary of State in Washington, March 31, 1972, National Archives and Records Administration.
25. "Important Memorandum from Turk NSC on Internal Situation," report from U.S. Embassy in Ankara to Secretary of State in Washington, DC, March 28, 1972, National Archives and Records Administration.
26. "Istanbul Politics: 1971 Highlights: 1972 Outlook" report from U.S. Consulate in Istanbul to the State Department in Washington, DC, January 18, 1972, National Archives and Records Administration.
27. Memo from the U.S. Consulate in Istanbul to the Secretary of State in Washington, DC, April 1972, National Archives and Records Administration.
28. "Ankara Martial Law Command Briefing on Terrorism in Turkey," from U.S. Embassy in Ankara to the Department of State, February 1, 1973, National Archives and Records Administration.

29. "Terrorist Group Profiles" report by the Vice President's Task Force on Combatting Terrorism, February 1986, from the National Security Archives.

30. CIA report "Turkey: Summary," no date (probably 1969), released 2002, CIA-RDP 76M00527R000700200001-1, CIA Research Tool (CREST), National Archives and Records Administration.

31. CIA report "Terrorism and the Fedayeen," September 1972, released 1999, CIA-RDP 79-01194A000200120001-1, CIA Research Tool (CREST), National Archives and Records Administration.

32. CIA report "International Terrorism in 1979," April 1980, released 2005, CIA-RDP 86-B00985R000200026002-1, CIA Research Tool (CREST), National Archives and Records Administration.

33. CIA report "West Europe Report," March 1981, released 2005, CIA-RDP 82-00850R000300030001-4, CIA Research Tool (CREST), National Archives and Records Administration.

LIST OF INTERVIEWS CONDUCTED IN TURKEY

1. Sengül Altan, Ankara, July 27, 2003.
2. Rehan Atasü/Ayşegul Oğuz at Flying Broom, Ankara, July 29, 2003.
3. Akin Atauz, former member of Turkish Labor Party, Ankara, June 30, 2006.
4. Ulus Baker, Istanbul, July 4, 2006.
5. General Edip Baser, Istanbul, July 25 2005.
6. Mihri Belli, former leader of the NDR faction of the Turkish Labor Party, Istanbul, June 21, 2006.
7. Faruk Bildirici, *Hürriyet*, Ankara, December 23, 2005.
8. Mehmet Birbiri, U.S.-Turkish liaison officer, İncirlik Thursday, July 2003.
9. Master Sergeant Blanco, İncirlik, July 2003.
10. Abdullah Çalışkan, AKP, June 9, 2005.
11. Cengiz Çandar, journalist and scholar, Istanbul, January 20, 2006.
12. Ilnur Cevik, editor at Turkish Daily News, Ankara, Monday, July 28, 2003.
13. Feride Ceyhan, Working Women's Association, Adana, August 5, 2003.
14. Sami Cohen, *Milliyet* newspaper, Istabul, June 23, 2006.
15. Nur Bilge Criss, professor at Bilkent University, Ankara, December 27, 2005.
16. Aydin Çubukçu, Ankara, December 26, 2005.
17. Kenan Durukan, President of Harb-İş 1971–1993, Istanbul, June 20, 2007.
18. Oktay Etiman, former member of THKP-C, Ankara, June 26, 2006.
19. Tom Evans, Family Advocacy Program Assistant at İncirlik, August 5, 2003.
20. Kaya Güvenç, former president of TMMOB, Ankara, June 9, 2005.

21. Fatoş Hacivelioğlu, Contemporary Lawyers Association / Çağdaş Hukukgular Derneği, Adana, August 6, 2003.
22. Mustafa Karaalioğlu, Yeni Şafak newspaper, Ankara, June 9, 2005.
23. Gülsen Kayır, Kadın Dayanışma Vakfı, Ankaralı Feministler, Ankara, June 9, 2005.
24. Ertuğurul Kürkçü, former head of Dev-Genç, now coordinator of the Independent Communications Network in Istanbul, July 4, 2006.
25. Masis Kürkçügil, activist since 1960s, Istanbul, summer 2006.
26. *Ömer* Laciner, activist since 1960s, currently editor of Birikim, Istanbul, summer 2006.
27. Mahmut, Ankara, June 27, 2005.
28. Tayfun Mater, Küresel BAK, Istanbul 8 June 2005.
29. Özhan Önder ODP, Istanbul, January 3, 2006.
30. Yıldız Önen, Küresel BAK, Istanbul, June 7, 2005.
31. Master Sargant Northcutt, İncirlik, July 2003.
32. Feray Salman, Secretary General of the Human Rights Association, Ankara, July 29, 2003.
33. Orhan Silier, Social and Economic History Foundation, Istanbul, August 9, 2003.
34. Nükhet Sirman, professor at Bogazici University in Istanbul, August 12, 2003.
35. Çetin Soyak, Vice-President of Harb-İş from 1971–1993, August 29, 2007.
36. Aslıhan Tümer, Greenpeace, Istanbul, June 27, 2005.
37. General Bahtiyar Türker, Ankara, December 25, 2005.
38. Vaughn, human relations advocate, İncirlik, July 2003.
39. Şiar Rişvanoğlu, Contemporary Lawyers Association / Çağdaş Hukukgular Derneği, Adana, August 6, 2003.
40. Faruk Ünsal, AKP, Ankara, June 9, 2005.
41. Ertuğrul Yalçınbayır, AKP, Ankara, June 9, 2005.
42. Murat Yetkin, *Radikal* newspaper, Ankara, June 9, 2005.
43. Mustafa Bulent Yoğun, Adana, August 4, 2003.
44. Interviews with various US military personnel stationed in Incirlik.

Bibliography

1961. "Military Bases of the USA and FRG in Europe." *International Affairs*: 108–110.

1980. "Turkey, Greece, and NATO: The Strained Alliance." Edited by Committee on Foreign Relations of the U.S. Senate.

1983. *Weißbuch der Bundeswehr*. Bonn: Bundesverteidigungsministerium der Bundesrepublik Deutschland.

1995. "United States Security Strategy for Europe and NATO." edited by Department of Defense. Washington, DC: Office of International Security Affairs.

2002. *Cumhuriyet Ansiklopedisi* Istanbul.

Abramowitz, Morton. 2000. *Turkey's Transformation and American Policy*. New York: Century Foundation Press.

Achilles, Olaf. 1987. *Tiefflug*: Bornheim: Lamuv Verlag.

Affairs, *Office of International Security*. 1995. "United States Security for Europe and NATO." Edited by Department of Defense: DoD, Washington, DC.

Ahmad, Feroz. 1977. *The Turkish Experiment in Democracy 1950–1975*. Boulder, CO: Westview Press.

———. 1993. *The Making of Modern Turkey*. London and New York: Routledge.

———. 2004. "The Historical Background of Turkey's Foreign Policy." In *The Future of Turkish Foreign Policy*, edited by Lenore G. Martin and Dimitris Keridis. Cambridge, MA: MIT Press.

Aldrich, Robert and John Connell. 1998. *The Last Colonies*. Cambridge: Cambridge University Press.

Alexander, Yonah and Dennis A. Pluchinsky. 1992. *Europe's Red Terrorists: The Fighting Communist Organizations*. London: Frank Cass.

Ali, Tariq. 2004. *Bush in Babylon: The Recolonisation of Iraq*. London: Verso.

Altınay, Ayşe Gül 2004. *The Myth of the Military-Nation: Militarism, Gender, and Education in Turkey*. New York: Palgrave.

Altınay, Ayşe Gül and Amy Holmes. 2009. "Resistance to the American Military Presence in Turkey in the Context of the Iraq War." In *The Bases of Empire: The Global Struggle against US Military Posts*, edited by Catherine Lutz: London: Pluto Press.

Alvah, Donna. 2007. *Unofficial Ambassadors: American Military Families Overseas and the Cold War 1946–1965*. New York: New York University Press.

Alvarez, David J. 1974. "The Missouri Visit to Turkey: An Alternative Perspective on Cold War Diplomacy." *Balkan Studies Journal* 15: 225–236.

Anderson, Jeffrey, John J. Ikenberry, and Thomas Risse (eds.). 2008. *The End of the West? Crisis and Change in the Atlantic Order*. Ithaca, NY: Cornell University.

Andries, Nathalie (ed.). 2004. *Zerreissprobe Frieden: Baden-Wuerttemberg und Der Nato-Doppelbeschlus*. Stuttgart: Haus der Geschichte.

Angerer, Jo, Sabine Lauxen, Erich Schmidt-Eenboom (eds.). 1990. *"Amerikanische Freunde": Die Politik der US-Streitkraefte in der Bundesrepublik*. Frankfurt: Sammlung Luchterhand Verlag.

Arrighi, Giovanni. 1994. *The Long Twentieth Century: Money, Power, and the Origins of Our Times*. London and New York: Verso.

2005. "Hegemony Unravelling – 2." *New Left Review* 33:83–117.

2007. *Adam Smith in Beijing: Lineages of the Twenty-First Century*. London and New York: Verso.

Arrighi, Giovanni, Beverly J. Silver, and Iftikhar Ahmad. 1999. *Chaos and Governance in the Modern World System*. Minneapolis: University of Minnesota Press.

Atkins, Sean. 2004. "What Influences Negative Domestic Reactions to Deploying Forces into Allied States." *Air Power History* 51 (Winter): 40–51.

Aydin, Mustafa. 2004. "Turkish Foreign Policy: Framework and Analysis." Ankara: Center for Strategic Research.

Bacevich, Andrew J. 2002. *American Empire: The Realities and Consequences of U.S. Diplomacy*. Cambridge, MA: Harvard University Press.

Bah, Umaru. 2008. "Daniel Lerner, Cold War Propaganda and US Development Communication Research: An Historical Critique." *Journal of Third World Studies* 25, Nr 1:183–198.

Bahro, Rudolf. 1982. *Socialism and Survival*. London: Heretic Books.

Baker, Anni P. 2004. *American Soldiers Overseas: The Global Military Presence*. London: Praeger.

Balfour-Paul, Glen. 1991. *The End of Empire in the Middle East: Britain's Relinquishment of Power in Her Last Three Arab Dependencies*. Cambridge and New York: Cambridge University Press.

Barreto, Amílcar Antonio. 2002. *Vieques, the Navy and Puerto Rican Politics*. Gainesville: University Press of Florida.

Bartholomew, Amy. 2006. *Empire's Law: The American Imperial Project and the "War to Remake the World."* London, Ann Arbor, MI, and Toronto: Pluto and Between the Lines.

Bebermeyer, Hartmut and Christian Thimann. 1989. *Die ökonomische Bedeutung der US-Streitkräfte in der Bundesrepublik. Eine Kosten-Nutzen-Analyse*. Köln: Deutscher Instituts-Verlag

Benhabib, Seyla. 1979. "The Next Iran or the Next Brazil: Right-wing Groups behind Political Violence in Turkey." *MERIP Reports* 77:16–17.

Berghan, Volker. 1981. *Militarism: The History of an International Debate 1861–1979*. Cambridge: Cambridge University Press.

Berktay, Halil. 1992. "The Search for the Peasant in Western and Turkish History/ Historiography." In *New Approaches to State and Peasant in Ottoman History*, edited by Suraiya Faroqhi, Halil Berktay. London: Frank Cass.

Betz, Hans-Georg. 1989. "Strange Love? How the Greens began to Love NATO." *German Studies Review* 12:487–505.

Bila, Fikret. 2004. *Sivil Darbe Girişimi ve Ankara'da Irak Savaşları*. Ankara: Ümit Yayıncılık.

Binnendijk, Hans., Alfred Friendly, and United States Congress Senate. Committee on Foreign Relations. 1980. *Turkey, Greece, and NATO: The Strained Alliance: A Staff Report to the Committee on Foreign Relations, United States Senate*. Washington: U.S. G.P.O.

Blackwill, Robert D. 1992. "Patterns of Partnership: The US-German Security Relationship in the 1990s." In *From Occupation to Cooperation: The United States and Germany in a Changing World Order*, edited by Steven Muller and Gebhard Schweigler. New York: W. W. Norton & Company.

Blaker, James R. 1990. *United States Overseas Basing: An Anatomy of the Dilemma*. New York: Praeger.

Boutwell, Jeffrey. 1990. *The German Nuclear Dilemma*. Ithaca, NY: Cornell University Press.

Bracken, Paul. 1995. "Reconsidering Civil-Military Relations." In *US Civil-Military Relations in Crisis or Transition?*, edited by Don M. Snider and Miranda A. Carlton-Crew. Washington, DC: The Center for Strategic and International Studies.

Bredthauer, Karl D. 1982. "Ziehen die Amerikaner ab?" *Blätter für deutsche und internationale Politik* 27.

Breyman, Steve. 1998. *Movement Genesis: Social Movement Theory and the 1980s West German Peace Movement*. Boulder, CO: Westview Press.

———. 2001. *Why Movements Matter: The West German Peace movement and U.S. Arms Control Policy*. Albany: State University of New York Press.

Burawoy, Michael. 1983. "Between the Labor Process and the State: the Changing Face of Factory Regimes under Advanced Capitalism." *American Sociological Review* 48.

Burbank, Jane and Cooper, Frederick. 2010. *Empires in World History: Power and the Politics of Difference*. Princeton: Princeton University Press.

Buro, Andreas. 1997. *Totgesagte Leben Länger: Die Friedensbewegung: Von Der Ost-West-Konfrontation Zur Zivilen Konfliktbearbeitung*. Idstein.

Butterwegge, Christoph, Klaus Jkubowski, Ekkehard Lentz, and Siegfried Ziegert (eds.). 1983. *Friedensbewegung – was nun? Probleme und Perspektiven nach der Raketenstationierung*. Hamburg: VSA-Verlag.

Byers, Michael. 2008. "Crisis, What Crisis? Transatlantic Differences and the Foundations of International Law." In *The End of the West? Crisis and Change in the Atlantic Order*, edited by Jeffrey Anderson, G. John Ikenberry, and Thomas Risse. Ithaca: Cornell University.

Cain, P. J., and A. G. Hopkins. 1993. *British Imperialism: Crisis and Deconstruction, 1914–1990*. London and New York: Longman.

Calder, Kent E. 2007. *Embattled Garrisons: Comparative Base Politics and American Globalism*. Princeton, NJ: Princeton University Press.

Caliskan, Koray, Yuksel Taskin. 2003. "Litmus Test: Turkey's Neo-Islamists Weigh War and Peace." *MERIP*. Published online Jan 30, http://www.koraycaliskan.net/docs/articles/K-Caliskan_Litmus_Test.pdf.

Calleo, David P. 1987. *Beyond American Hegemony: The Future of the Western Alliance*. New York: Basic Books.

1992. *The Bankrupting of America: How the Federal Budget is Impoverishing the Nation*. New York: W. Morrow.

Candar, Cengiz. 2000. "Some Turkish Perspectives on the United States and American Policy toward Turkey." In *Turkey's Transformation and American Policy*, edited by Morton Abramowitz. New York: Century Foundation Press.

Cartwright, John and Julian Critchley. 1985. *Cruise, Pershing and SS-20: The Search for Consensus: Nuclear Weapons in Europe*. London: Brassey's Defense Publishers.

Castellon, David. 1998. "Candy Bomber Delivers Happiness." *Air Force Times* 58.

Chandler, William M. and Alan Siaroff. 1986. "Postindustrial Politics in Germany and the Origins of the Greens." *Comparative Politics* 18:303–325.

Cleveland, William L. 2000. *A History of the Modern Middle East*. Boulder, CO: Westview Press.

Cohen, Stephen S., and John Zysman. 1987. *Manufacturing Matters: The Myth of the Post-Industrial Economy*. New York: Basic Books.

Cook, Schura. 1982. "Germany: From Protest to Terrorism." In *Terrorism in Europe*, edited by Yonah Alexander and Kenneth A. Myers. New York: St. Martin's Press – Center for Strategic and International Studies Georgetown.

Cook, Steven A. 2007. *Ruling but Not Governing: The Military and Political Development in Egypt, Algeria, and Turkey*. Baltimore, MD: Johns Hopkins University Press.

Cook, Steven A., and Elizabeth Sherwood-Randall. 2006. "Generating Momentum for a New Era in US-Turkey Relations." New York: Council on Foreign Relations.

Cooley, Alexander. 2008. *Base Politics: Democratic Change and the U.S. Military Overseas*. Ithaca, NY: Cornell University Press.

Cooper, Alice Holmes. 1996. *Paradoxes of Peace: German Peace Movements since 1945*. Ann Arbor: University of Michigan Press.

Cortright, David. 2006. "The World Says No: The Global Movement against War in Iraq." In *Iraq Crisis and World Order: Structural, Institutional, and Normative Challenges*, edited by Ramesh Chandra Thakur and Waheguru Pal Singh Sidhu. Tokyo: United Nations University Press.

Cossaboom, Robert and Gary Leiser. 1998. "Adana Station 1943–45: Prelude to the Post-War American Military Presence in Turkey." *Middle Eastern Studies* 34:73–86.

Cottrell, Alvin J. and Thomas H. Moorer. 1977. *U.S. Overseas Bases: Problems of Projecting American Military Power Abroad*. London: Sage Publications.

Criss, Nur Bilge. 1993. "US Forces in Turkey." In *US Military Forces in Europe: The Early Years 1945–1970*, edited by Simon W. Duke and Wolfgang Krieger. Boulder, CO: Westview Press.

1999. *Istanbul under Allied Occupation 1918–1923*. Leiden: Brill.

2002. "A Short History of Anti-Americanism and Terrorism: The Turkish Case." *The Journal of American History* 89:472–484.

Crossley, Nick. 2002. *Making Sense of Social Movements*. Philadelphia, PA: Open University Press.

CSIA European Security Working Group. 1978/79. "Instability and change on NATO's Southern Flank." *International Security* 3.

Cunningham, Keith B. and Andreas Klemmer. 1995. "Restructuring the U.S. Military Bases in Germany: Scope, Impacts, and Opportunities." Bonn International Center for Conversion.

Czempiel, Ernst Otto. 1989. *Machtprobe: Die USA Und Die Sowjetunion in Den Achtziger Jahren: C. H. Beck, 1989*. München: C. H. Beck.

Daalder, Ivo. 2003. "The End of Atlanticism." The Brookings Institution.

Darwin, John. 1988. *Britain and Decolonisation: The Retreat from Empire in the Post-War World*. New York: St. Martin's Press.

Davidson, Eugene. 1959. *The Death and Life of Germany*. New York: Knopf.

Deiseroth, Dieter. 1988. "US-Basen in der Bundesrepublik: Stationierungsrechte und Interventionen ausserhalb des NATO-Gebietes." Weilheim: Forschungsinstituts für Friedenspolitik Starnberg.

Della Porta, Donnatella. 1995. *Social Movements, Political Violence, and the State: A Comparative Analysis of Italy and Germany*. Cambridge: Cambridge University Press.

Della Porta, Donatella and Rucht, Dieter. 1995. "Social Movement Sectors in Context: A Comparison of Italy and West Germany, 1965–1990," in J. Craig Jenkins e Bert Klandermans (eds.), *The Politics of Social Protest*, Minneapolis, Minnesota University Press, 299–272.

De Nardo, James. 1985. *Power in Numbers: The Political Strategy of Protest and Rebellion*. Princeton, NJ: Princeton University Press.

Department, Defense. 2001. "Quadrennial Defense Review Report." Washington, DC: Department of Defense.

Department, State. "Foreign Relations of the United States, 1961–1963, Volume XVI, Eastern Europe; Cyprus; Greece; Turkey." Edited by State Department: United States Government Printing Office.

1993. "Foreign Relations of the United States, 1958–1960 Volume X Part 2: Eastern Europe; Finland; Greece; Turkey." Edited by State Department: United States Government Printing Office.

Deppe, Frank. 2000. "Die Linke in der Geschichte der Bundesrepublik Deutschland." Hamburg: Sozialismus – VSA Verlag.

Deringil, Selim. 1989. *Turkish Foreign Policy during the Second World War: An Active Neutrality*. Cambridge and New York: Cambridge University Press.

Dietrich, Barbara and Erich Schmidt-Eenboom. 1987. *Der militarisierte Frieden: Studie zur Militarisierung der Rhein-Main Region*. Starnberg: Forschungsinstitut fuer Friedenspolitik.

Diner, Dan. 1983. "The "National Question" in the Peace Movement – Origins and Tendencies." *New German Critique* Winter:86–107.

1996. *America in the Eyes of Germans: An Essay on Anti-Americanism* Princeton, NJ: Markus Wiener Publishers.

Docena, Herbert. 2007. "At the Door of All the East: The Philippines in United States Military Strategy." Quezon City: Focus on the Global South.

Doyle, Michael W. 1986. *Empires*. Ithaca, NY: Cornell University Press.

Dubbe, Daniel., and Thorwald Proll. 2003. *Wir kamen vom anderen Stern: Über 1968, Andreas Baader und ein Kaufhaus*. Edition Nautilus.

Duke, Simon. 1993. *The Burdensharing Debate: A Reassessment*. New York: St. Martin's Press.

Duke, Simon, and Stockholm International Peace Research Institute. 1989. *United States Military Forces and Installations in Europe*. Solna, Sweden Oxford and New York: SIPRI and Oxford University Press.

Duke, Simon and Wolfgang Krieger. 1993. *U.S. Military Forces in Europe: The Early Years, 1945–1970*. Boulder, CO: Westview Press.

Dur, Lt Comm P. A. 1974. "The U.S. Sixth Fleet: Search for Consensus." *United States Naval Institute* 100.

Enloe, Cynthia. 2000. *Bananas, Beaches, and Bases: Making Feminist Sense of International Politics*. Berkeley: University of California Press.

Evriviades, Marios L. 1984. "The Evolving Role of Turkey in U.S. Contingency Planning and Soviet Reaction." John F. Kennedy School of Government.

1998. "Turkey's Role in the United States Strategy during and after the Cold War." *Mediterranean Quarterly* 9:22.

Fanon, Frantz. 1963. *The Wretched of the Earth*. New York: Grove Press.

Faroqhi, Suraiya, and Halil Berktay. 1992. *New Approaches to State and Peasant in Ottoman History*. London: Frank Cass.

Forum, Initiative Sozialistisches (ed.). 1984. *Frieden: je näher man hinschaut desto fremder schaut es zurück*. Freiburg: Ça-Ira Verlag.

Foster, John Bellamy, and Robert Waterman McChesney. 2004. *Pox Americana: Exposing the American Empire*. New York: Monthly Review Press.

Franck, Thomas. 1995. *Fairness in International Law and Institutions*. Oxford: Clarendon Press.

Fülberth, Georg. 1999. *Berlin-Bonn-Berlin: Deutsche Geschichte seit 1945*. Köln: Papy Rossa Verlag.

2007. *Finis Germaniae: Deutsche Geschichte seit 1945*. Köln: PapyRossa Verlag.

Gaddis, John Lewis. 2005. *Strategies of Containment: A Critical Appraisal of American National Security Policy During the Cold War*. Oxford: Oxford University Press.

Gamson, William. 1975. *The Strategy of Social Protest*. Homewood, Ill: Dorsey Press.

Ganser, Daniele. 2005. *NATO's Secret Armies: Operation Gladio and Terrorism in Western Europe*. London: Frank Cass.

Garnham, David. 1994. "Ending Europe's Security Dependence," *Journal of Strategic Studies* 17, no. 4, December.

Geiss, Imanuel. 1994. "Great Powers and Empires: Historical Mechanisms of their Making and Breaking." In *The fall of Great Powers: Peace, Stability, and Legitimacy*, edited by Geir Lundestad. Oxford: Oxford University Press.

Gerson, Joseph and Bruce Birchard (eds.). 1991. *The Sun Never Sets... Confronting the Network of Foreign U.S. Military Bases*. Boston, MA: South End Press.

Gillem, Mark L. 2007. *America Town: Building the Outposts of Empire*. Minneapolis: University of Minnesota Press.

Gilpen, Robert G. 1984. "The Dual Problems of Peace and National Security." *PS* 17:18–23.

Gimbel, John. 1968. *The American Occupation of Germany; Politics and the Military, 1945–1949*. Stanford, CA: Stanford University Press.

Giugni, Marco, Doug McAdam and Charles Tilly (eds.). 1999. *How Social Movements Matter*. Minneapolis: University of Minnesota Press.

Giugni, Marco., Doug McAdam, and Charles Tilly. 1998. *From Contention to Democracy*. Lanham, MD: Rowman & Littlefield Publishers.

Glaser, Daniel. 1946. "The Sentiments of American Soldiers Abroad toward Europeans." *American Journal of Sociology* 51:433–438.

Goedde, Petra. 2003. *GIs and Germans: Culture, Gender and Foreign Relations, 1945–1949*. London: Yale University Press.

Göl, Ayla. 2006. "Iraq and World Order: A Turkish Perspective." In *Iraq Crisis and World Order: Structural, Institutional, and Normative Challenges*, edited by Ramesh Chandra Thakur and Waheguru Pal Singh Sidhu. Tokyo: United Nations University Press.

Gonlubol, Mehmet. 1975. "NATO and Turkey." In *Turkey's Foreign Policy in Transition 1950–1974*, edited by Kemal Karpat. Leiden: E. J. Brill.

Goodwin, Jeff. 2006. "A Theory of Categorical Terrorism." *Social Forces* 84.

Gordon, Philip H., and Jeremy Shapiro. 2004. *Allies at war America, Europe, and the crisis over Iraq*. New York: McGraw-Hill.

Gramsci, Antonio, Joseph A. Buttigieg, and Antonio Callari. 1991. *Prison Notebooks*. New York: Columbia University Press.

Gress, David. 1985. *Peace and Survival: West Germany, the Peace Movement and European Security*. Stanford: Hoover Press Publication.

Gress, Denis L. and David R. Bark. 1989. *Democracy and Its Discontents*. Oxford: Basil Blackwell.

Grossmann, Atina. 2007. *Jews, Germans, and Allies: Close Encounters in Occupied Germany*. Princeton, NJ: Princeton University Press.

Güney, Aylin. 2005. "An Anatomy of the Transformation of the US-Turkish Alliance: From 'Cold War' to War on Iraq'" *Turkish Studies* 6:341–359.

Gürkan, Ihsan. 1980. *NATO, Turkey, and the Southern Flank: A Mideastern Perspective*. London: National Strategy Information Center.

Gurr, Ted. 1970. *Why Men Rebel*. Princeton, NJ: Princeton University Press.

Güvenç, Serpil. 2005. "Socialist Perspectives on Foreign Policy Issues: The Case of TIP in the 1960s." Ankara: Middle East Technical University.

Habermas, Jürgen, and Ciaran Cronin. 2006. *The Divided West*. Cambridge, UK and Malden, MA: Polity.

Hakki, Murat Metin. 2007. *The Cyprus Issue: A Documentary History, 1878–2006*. London: I. B. Tauris.

Hale, William. 1994. *Turkish Politics and the Military*. London: Routledge.

——— 1999. "Foreign Policy and Domestic Politics." In *The Turkish Republic at 75 Years: Progress – Development – Change*, edited by David Shankland: Eothen Press.

——— 2000. *Turkish Foreign Policy 1774–2000*. London: Frank Cass.

——— 2007. *Turkey, the US, and Iraq*. London: London Middle East Institute.

Halliday, Fred. 1986. *The Making of the Second Cold War*. London: Verso.

Hardt, Michael and Antonio Negri. 2004. *Multitude: War and Democracy in the Age of Empire*. New York: Penguin Press.

——— 2000. *Empire*. Cambridge: Harvard University Press.

Harkavy, Robert E. 1982. *Great Power Competition for Overseas Bases: The Geopolitics of Access Diplomacy*. New York: Pergamon Press.

——— 1989. *Bases Abroad: The Global Foreign Military Presence*. Oxford and New York: Oxford University Press.

Harrington, Daniel F. 1998. "The Air Force Can Deliver Anything! A History of the Berlin Airlift." Edited by USAFE Office of History: Ramstein AFB.

Harris, George. 1972. *Troubled Alliance: Turkish-American Problems in Historical Perspective, 1945–1971*. Washington DC: American Enterprise Institute.

Harris, George S. 2002. *The Communists and the Kadro Movement: Shaping Ideology in Atatürk's Turkey*. Istanbul: The ISIS Pr.

Harvey, David. 2005. *The New Imperialism*. Oxford and New York: Oxford University Press.

Haumann, Wilhelm and Thomas Petersen. 2004. "German Public Opinion on the Iraq Conflict: A Passing Crisis with the USA or a Lasting Departure?" *International Journal of Public Opinion Research* 16:311–330.

Hawkins, John P. 2001. *Army of Hope, Army of Alienation: Culture and Contradiction in the American Army Communities of Cold War Germany*. Westport: Praeger.

Hawkins, John Palmer. 2005. *Army of Hope, Army of Alienation: Culture and Contradiction in the American Army Communities of Cold War Germany*. Tuscaloosa: University of Alabama Press.

Herf, Jeffrey. 1991. *War by Other Means: Soviet Power, West German Resistance, and the Battle of the Euromissiles*. New York: The Free Press.

Hickok, Michael Robert. 2000. "Hegemon Rising: The Gap Between Turkish Strategy and Military Modernization." *Parameters: US Army War College* 30:105–120.

Ho, Engseng. 2004. "Empire through Diasporic Eyes: A View from the Other Boat." *Society for Comparative Study of Society and History* Volume 46, 2: 210–246.

Hobsbawm, E. J. 1994. *The Age of Extremes: A History of the World, 1914–1991*. New York: Pantheon Books.

Hockenos, Paul. 2008. *Joschka Fischer and the Making of the Berlin Republic: An Alternative History of Postwar Germany*. Oxford and New York: Oxford University Press.

Hoffmann, Martin 1997. *Rote Armee Fraktion: Texte und Materialien zur Geschichte der RAF*. Berlin ID-Verlag.

Höhn, Maria. 2002. *GIs and Fräuleins: the German-American Encounter in 1950s West Germany*. Chapel Hill: University of North Carolina Press.

Höhn, M., and Klimke, M. 2010. *A Breath of Freedom: The Civil Rights Struggle, African American GIs, and Germany*. New York: Palgrave Macmillan.

Holland, R. F. 1985. *European Decolonization, 1918–1981: An Introductory Survey*. New York: St. Martin's Press.

Holmes, Kim R. 1984. *The West German Peace Movement and the National Question*. Cambridge, MA: Institute for Foreign Policy Analysis.

Hoopes, Townsend. 1958. "Overseas Bases in American Strategy." *Foreign Affairs* October 1, 1958. Accessed January 2, 2014. http://www.foreignaffairs.com/articles/71439/townsend-hoopes/overseas-bases-in-american-strategy.

Horchem, Hans Joseph. 1991. "The Terrorist Lobby in West Germany: Campaigns and Propaganda in Support of Terrorism." In *Tolerating Terrorism in the West: An International Survey*, edited by Noemi Gal-Or. London: Routledge.

Howard, Michael. 1982/83. "Reassurance and Deterrence: Western Defense in the 1980s." *Foreign Affairs* 61.

Howe, Stephen. 1993. *Anticolonialism in British Politics: The Left and the End of Empire, 1918–1964*. New York: Oxford University Press.

Huntington, Samuel. 1972. *The Soldier and the State*. Cambridge: The Belknap Press.

Huston, James A. 1988. *Outposts and Allies: U.S. Army Logistics in the Cold War, 1945–1953*. London: Associated University Presses.

Iggers, Georg G., Q. Edward Wang, and Supriya Mukherjee. 2008. *A global history of modern historiography*. Harlow, England: Pearson Longman.

Isenberg, David. 2004. "The New US Global Posture Review: Reshaping America's Global Military Footprint." Asia Times online Aug 20. http://www.atimes.com/atimes/Front_Page/FH20Aa01.html.

Janowitz, Morris. 1960. *The Professional Soldier: A Social and Political Portrait*. Glencoe, IL: The Free Press.

Joffe, Josef. 1987. "Peace and Populism: Why the European Anti-Nuclear Movement Failed." *International Security* 11:3–40.

Johnson, Chalmers A. (ed.). 1999. *Okinawa: Cold War Island*. Cardiff, CA: Japan Policy Research Institute.

2000. *Blowback: The Costs and Consequences of American Empire.* New York: Metropolitan Books.

2004. *The Sorrows of Empire: Militarism, Secrecy, and the End of the Republic.* New York: Metropolitan Books.

2006. *Nemesis: The Last Days of the American Republic.* New York: Metropolitan Books.

2010. *Dismantling the Empire: America's Last Best Hope.* New York: Metropolitan Books.

Johnstone, Diana. 1984. *The Politics of Euromissiles: Europe's Role in America's World.* London: Verso.

Jung, Dietrich and Wolfgango Piccoli. 2001. *Turkey at the Crossroads: Ottoman Legacies and a Greater Middle East.* London: Zed Books.

Kagan, Robert. 2003. *Of Paradise and Power: America and Europe in the New World Order*: New York: Knopf.

2004. "America's Crisis of Legitimacy." *Foreign Affairs* 83:65–87.

Kane, Tim. 2004. "Global US Troop Deployment 1950–2003." Heritage Foundation.

Kaplan, Lawrence S., Robert W. Clawson and Raimondo Luraghi (eds.). 1985. *NATO and the Mediterranean.* Wilmington, DE: Scholarly Resources Inc.

Karapin, Roger. 2007. *Protest Politics in Germany: Movements on the Left and Right since the 1960s.* University Park: Pennsylvania State University Press.

Karasapan, Omer. 1989. "Turkey and US Strategy in the Age of Glasnost." *Middle East Report* 160:4–10.

Karpat, Kemal. 1967. "Socialism and the Labor Party of Turkey." *Middle East Journal* 21.

(ed.). 1975. *Turkey's Foreign Policy in Transition 1950–1974.* Leiden: Brill.

Katzenstein, Peter. 2002. "Same War, different views: Germany, Japan, and the War on Terrorism." *Current History* 101.

2005. *A World of Regions: Asia and Europe in the American Imperium.* Ithaca: Cornell University Press.

Katzenstein, Peter. J., and Robert O. Keohane (eds). 2007. *Anti-Americanisms in World Politics.* Ithaca: Cornell University Press.

Kennedy, Paul M. 1987. *The Rise and Fall of the Great Powers: Economic Change and Military Conflict from 1500 to 2000.* New York: Random House.

Keyder, Caglar. 1979. "The Political Economy of Turkish Democracy." *New Left Review* 28:3–44.

2004. "The Turkish Bell Jar." *New Left Review*:65–84.

Kitfield, James. 2005a. *War & Destiny: How the Bush Revolution in Foreign and Military Affairs Redefined American Power.* Washington, DC: Potomac Books.

2005b. "Over There: An Independent Commission Raised Questions about a Pentagon Plan to Move Troops Based Overseas to the United States. Was anyone listening?" *National Journal* Volume 37, 43 (2005):3268.

Kitschelt, Herbert P. 1986. "Political Opportunity Structures and Political Protest: Anti-Nuclear Movements in Four Democracies." *British Journal of Political Science* 16:57–85.

Klein, Hans-Joachim. 1980. *La mort mercenaire. Témoignage d'un ancient terroriste Ouest-Allemande.* Paris: Editions du Seuil.

Koch, Uwe. 1983. "Friedenskampf ist Klassenkampf!" In *Die Friedensbewegung nach der Raketenstationierung,* edited by Christoph U. A. Butterwegge. Hamburg: VSA.

Koch, Connie., and Sauermann, Barbara. 2003. *2/15: The Day the World Said No to War*. Oakland, CA: AK Press.

Kolb, Felix. 2007. *Protest and Opportunities: Political Outcomes of Social Movements*. Frankfurt: Campus Verlag.

Koopmans, Ruud. 1993. "The Dynamics of Protest Waves: West Germany, 1965–1989." *American Sociological Review* 58:637–658.

Kraushaar, Wolfgang., Jan Philipp Reemtsma, and Karin Wieland. 2005. *Rudi Dutschke, Andreas Baader, und die RAF*. Hamburg: Hamburger Edition.

Krepinevich, Andrew. 2003. "Meeting the Anti-Access and Area-Denial Challenge." Washington, DC: Center for Strategic and Budgetary Assessments.

Kuniholm, Bruce R. 1980. *The Origins of the Cold War in the Near East: Great Power Conflict and Diplomacy in Iran, Turkey, and Greece*. Princeton, NJ: Princeton University Press.

——— 1983. "Turkey and NATO: past, present, and future". *Orbis* 272:421–445.

Lafontaine, Oskar. 1983. *Angst vor den Freunden: die Atomwaffenstrategie der Supermaechte zerstoert die Bündnisse*. Hamburg: Rowohlt.

Landau, Jacob M. 1974. *Radical Politics in Modern Turkey*. Leiden: E. J. Brill.

Langguth, Gerd. 1983. *Protestbewegung – Niedergang – Renaissance: Die Neue Linke seit 1968*. Köln: Verlag Wissenschaft und Politik.

Lawson, Fred. 2004. "Political Economy, Geopolitics and the Expanding US Military Presence in the Persian Gulf and Central Asia." *Critical Middle Eastern Studies* 13:7–31.

Leffler, Melvyn P. 1992. "A Preponderance of Power National Security, the Truman Administration, and the Cold War." In *Stanford Nuclear Age Series*. Stanford, CA: Stanford University Press.

Lerner, Daniel. 1958. *The Passing of Traditional Society: Modernizing the Middle East*. Glencoe, IL: Free Press.

Lesser, Ian O. 2006. "Turkey, the United States and the Delusion of Geopolitics." *Survival* 48:83–96.

Lewis, Jesse W. 1976. *The Strategic Balance in the Mediterranean*. Washington, DC: American Enterprise Institute for Public Policy Research.

Lipovsky, Igor P. 1992. *The Socialist Movement in Turkey, 1960–1980*. Leiden and New York: E. J. Brill.

Livingston, Craig. 1994. "'One thousand wings': the United States Air Force Group and the American Mission for Aid to Turkey, 1947–50." *Middle Eastern Studies* Volume 30, 48:778–825.

Lovatt, Debbie. 2001. *Turkey Since 1970: Politics, Economics and Society*. New York: Palgrave.

Luber, Burkhard. 1986. *Militaeratlas von Flensburg bis Dresden*. Bonn: Die Gruenen.

Lundestad, Geir. 1994. *The Fall of Great Powers: Peace, Stability, and Legitimacy*. Oslo Oxford and New York: Scandinavian University Press and Oxford University Press.

——— 2003. *The United States and Western Europe since 1945: From "Empire" by Invitation to Transatlantic Drift*. Oxford and New York: Oxford University Press.

Lutz, Catherine. 2001. *Homefront: A Military City and the American Twentieth Century*. Boston, MA: Beacon Press.

——— 2006. "Empire is in the Details." *American Ethnologist* 33:593–611.

——— (ed.). 2009. *The Bases of Empire: The Global Struggle against US Military Posts*. London: Pluto Press.

Lyons, Matthew. 1999. "The Grassroots Movement in Germany, 1972–1985." In *Nonviolent Social Movements: A Geographical Perspective*, edited by Lester R. Kurtz Stephen Zunes, Sarah Beth Asher. Oxford: Blackwell Publisher.

Mackenzie, Kenneth. 1984. *Turkey in Transition: The West's Neglected Ally*. London: Institute for European Defense and Strategic Studies.

Mahncke, Dieter (ed.). 1991. *Amerikaner in Deutschland: Grundlagen und Bedingungen der transatlantischen Sicherheit*. Bonn: Bouvier Verlag.

Maier, Charles S. 2002. "An American Empire? The Problems of Frontiers and Peace in Twenty-First Century World Politics." *Harvard Magazine*, 105, 2 (November-December).

Mango, Andrew. 2004. *The Turks Today*. London: John Muray.

2005. *Turkey and the War on Terror: For Forty Years We Fought Alone*. London: Routledge.

Mann, Michael. 2003. *Incoherent Empire*. London and New York: Verso.

Mardin, Serif. 1997. "Projects as Methodology: Some Thoughts on Modern Turkish Social Science." In *Rethinking Modernity & National Identity in Turkey*, edited by Sibel Bozdoğan and Kasaba Bozdoğan, Seattle: University of Washington Press.

Marin, Lou. 2007. "The German Peace Movement Confronts the US Military." Unpublished Manuscript.

Markovits, Andrei and Philip S. Gorski. 1993. *The German Left: Red, Green, and Beyond*. Cambridge: Polity Press.

Mayall, Simon V. 1977. "Turkey: Thwarted Ambition." Washington, DC: Institute of National Strategic Studies, National Defense University.

Mayer, Margit and John Ely (eds.). 1998. *The German Greens: Paradox between Movement and Party*. Philadelphia, PA: Temple University Press.

McCaffrey, Katherine T. 2002. *Military Power and Popular Protest: the U.S. Navy in Vieques, Puerto Rico*. New Brunswick, NJ: Rutgers University Press.

McCormick, Thomas J. 1995. *America's Half-Century: United States Foreign Policy in the Cold War and After*. Baltimore, MD: The Johns Hopkins University Press.

McDonald, John W. and Diane B. Bendahmane. 1990. *U.S. Bases Overseas: Negotiations with Spain, Greece, and the Philippines*. Boulder, CO: Westview Press.

McGhee, George. 1990. *The US-Turkish-NATO Middle East Connection: How the Truman Doctrine Contained the Soviets in the Middle East*. New York: St. Martin's Press.

McIntyre, W. David. 1998. *British Decolonization, 1946–1997: When, Why, and How Did the British Empire Fall?* New York: St. Martin's Press.

McNeill, William H. 1983. *The Pursuit of Power: Technology, Armed Force, and Society since A.D. 1000*. Oxford: Basil Blackwell.

Mechtersheimer, Afred. 1984. *Zeitbombe NATO*. Diederichs.

Mello, Brian. 2008. "Alternative Collective Subjectivities and the Political Impact of Radical Labor Activism in Japan and Turkey." In *Annual Meeting of the Midwest Political Science Association*.

Melman, Seymour. 1974. *The Permanent War Economy; American Capitalism in Decline*. New York: Simon and Schuster.

Migdalovitz, Carol. 2002. "Iraq: The Turkish Factor CRS Report for Congress." Washington, DC: Congressional Research Service.

Moeller, Benton G. 1995. "Learning from Each Other: The American Forces as Employers." In *Neighbor America: Americans in Rhineland-Palatinate, 1945–1995*, edited by Winfried Herget. Trier: Wissenschaftlicher Verlag Trier.

Moon, Katharine H. S. 1997. *Sex among Allies: Military Prostitution in U.S.-Korea Relations*. New York: Columbia University Press.

Müller, Emil-Peter. 1986. *Antiamerikanismus in Deutschland: Zwischen Care-Paket und Cruise Missile*. Koln: Deutscher Instituts-Verlag.

Mumcu, Uğur, and Mehmet Ali Aybar. 1986. *Aybar ile Söyleşi: Sosyalizm ve Bağımsızlık*. İstanbul: Tekin Yayınevi.

Mushaben, Joyce Marie. 1986. "Grassroots and Gewaltfreie Aktionen: A Study of Mass Mobilization Strategies in the West German Peace Movement." *Journal of Peace Research* 23:141–154.

Nabulsi, Karma. 1999. *Traditions of War: Occupation, Resistance, and the Law*. Oxford: Oxford University Press.

Nathan, James A. 1975. "A Fragile Detente: the U-2 Incident Re-examined." *Military Affairs* 39:97–104.

Nelson, Daniel J. 1987. *Defenders or Intruders?: The Dilemmas of U.S. Forces in Germany*. Boulder, CO: Westview Press.

——— 1987. *A History of US Military Forces in Germany*. Boulder, CO: Westview Press.

Nick, Volker, Volker Scheub, and Christof Then. 1993. *Mutlangen 1983–1987: Die Stationierung Der Pershing II und die Kampagne Ziviler Ungehorsam bis zur Abrüstung*: Eigenverlag. Künstlerei im Sudhaus: Tübingen.

Nitze, P. H., Rearden, S. L., and Smith, A. M. 1. 1989. *From Hiroshima to Glasnost: At the Center of Decision: A Memoir*. New York: Grove Weidenfeld.

O'Brien, Patrick Karl, and Armand Cleese. 2002. *Two Hegemonies: Britain 1846–1914 and the United States 1941–2001*. Aldershot, Hants, England and Burlington, VT: Ashgate.

O'Hanlon, Michael. 2008. "Unfinished Business: U.S. Overseas Military Presence in the 21st Century." Washington, DC: Brookings Institution.

Ohlemacher, Thomas. 1993. *Brücken der Mobilisierung: Soziale Relais und persönliche Netzwerke in Bürgerinitiativen gegen militärischen Tiefflug*. Wiesbaden: Deutscher Universitäts Verlag.

Önis, Ziya and Suhnaz Yilmaz. 2005. "The Turkey-EU-US Triangle in Perspective: Transformation or Continuity." *Middle East Journal* 59.

Owen, Edward Roger John. 2004. *State, Power and Politics in the Making of the Modern Middle East*. London and New York: Routledge.

Pfaltzgraff, Robert L. 1983. "The Greens of West Germany: Origins, Strategies, and Transatlantic Implications." Washington, DC: Institute for Foreign Policy Analysis.

Pflieger, Klaus. 2004. *Die Rote Armee Fraktion-RAF: 14.5.1970 bis 20.4.1998*. Baden-Baden: Nomos.

Pflüger, Tobias. 2006. "Es war gut, ihnen nicht geglaubt zu haben: Die deutsche Beteiligung am Irak-Krieg." *ak – zeitung für linke debate und praxis* Nr 504.

Pilat, J. F. 1980. *Ecological Politics: The Rise of the Green Party*. Georgetown University. Center for Strategic and International Studies, Thousand Oaks, CA: Sage Publications.

Piven, Frances and Richard Cloward. 1977. *Poor People's Movements: Why they Succeed, How they Fail*. New York: Pantheon Books.

Pluchinsky, Dennis A., and Yonah Alexander. 1992. *European Terrorism: Today & Tomorrow*. Washington: Brassey's.

Pohrt, Wolfgang. 1982. *Endstation: ueber die Wiedergeburt Der Nation*. Berlin: Rotbuch Verlag.

Pollard, Robert A. 1985. *Economic Security and the Origins of the Cold War, 1945–1950*. New York: Columbia University Press.

Pope, Nicole and Hugh Pope. 1997. *Turkey Unveiled: Atatürk and After*. London: John Murray.

2004. *Turkey Unveiled: A History of Modern Turkey*. Woodstock, NY: The Overlook Press.

Rabert, Bernhard. 1995. *Links- und Rechtsterrorismus in der Bundesrepublik Deutschland von 1970 bis heute*. Bonn: Bernard & Graefe Verlag.

Raschke, Joachim (ed.). 1993. *Die Grünen: Wie sie wurden, was sie sind*. Köln: Büchergilde Gutenberg.

Ricardo, Rogert. 1994. *Guantanamo – The Bay of Discord: The Story of the US Military Base in Cuba*. Melbourne: Ocean Press.

Ritzenhofen, Ute. 1995. "Working for the Americans: German-American Labor Relations." In *Neighbor America: Americans in Rhineland-Palatinate, 1945–1995*, edited by Winfried Herget. Trier: Wissenschaftlicher Verlag Trier.

Robins, Philip. 2003. "Confusion at Home, Confusion Abroad: Turkey between Copenhagen and Iraq." *International Affairs* 79, 3, May: 547–566.

Louis, W.R. and Robinson, R. 1994. "The Imperialism of Decolonization." *JICH* 22:462–511.

Rogers, Bernard General. 1982. "The Atlantic Alliance: Prescriptions for Difficult Decade." *Foreign Affairs* 60:1145–1156.

Rohrschneider, Robert. 1993. "Impact of Social Movements on European Party Systems." *Annals of the American Academy of Political and Social Science* 528:157–170.

Rubin, Barry. 2002. *Istanbul Intrigues*. Istanbul: Boğaziçi University Press.

Rucht, Dieter. 2003. "The Changing Role of Political Protest Movements." *West European Politics* 26:153–176.

Rucht, Dieter and Stefaan Walgrave (eds.). 2008. *Protest Politics: Demonstrations against the War on Iraq in the US and Western Europe*. Minneapolis: University of Minnesota Press.

2010. *The World Says No to War: Demonstrations against the War on Iraq* Minneapolis: University of Minnesota Press.

Rustow, Dankwart A. 1987. *Turkey: America's Forgotten Ally*. New York: Council of Foreign Relations.

Sakallioğlu, Ümit Cizre. 1997. "The Anatomy of the Turkish Military's Political Autonomy." *Comparative Politics* 29:151–166.

Salomon, Kim. 1986. "The Peace Movement: An Anti-Establishment Movement." *Journal of Peace Research* 23:115–127.

Samim, Ahmet. 1981. "The Tragedy of the Turkish Left." *New Left Review*. I/126:60–85.

Sandars, C. T. 2000. *America's Overseas Garrisons: the leasehold empire*. Oxford: Oxford University Press.

Scahill, Jeremy. 2007. *Blackwater: The Rise of the World's Most Powerful Mercenary Army*. New York, NY: Nation Books.

Schake, Kurt Wayne. 1998. *Strategic Frontier: American Bomber Command Bases Overseas 1950–1960*. Trondheim: Historisk Institutt

Schlau, Wilfried (ed.). 1990. *Allierte Truppen in der Bundesrepublik Deutschland*. Bonn: Köllen-Druck.

Schmidt-Eenboom, Erich and Jo Angerer (eds.). 1993. *Siegermacht NATO: Dachverband der neuen Weltordnung*. Muenchen: AK-Druck.

Schraut, Hans-Jürgen. 1993. "US Forces in Germany 1945–1955." In *US Military Forces in Europe, 1945–70*, edited by Simon and Wolfgang Krieger Duke. Boulder, CO: Westview Press.

Schultz, Nikolaus. 2006. "Was the War on Iraq illegal? The German Federal Administrative Court's Ruling of 21st June 2005." *German Law Journal* 7.

Seiler, Signe. 1985. *Die GIs: Amerikanische Soldaten in Deutschland*. Hamburg: Rowohlt.

Sever, Ayşegül. 2002. "Turkey and US On Iraq Since the Gulf War." *Turkish Review of Middle East Studies* 13.

Sharp, Jeremy M. and Christopher M. Blanchard. 2005. "Post-War Iraq: A Table and Chronology of Foreign Contributions." In *CRS Report for Congress*: The Library of Congress.

Silver, Beverly J. 2003. *Forces of Labor: Workers' Movements and Globalization since 1870*. Cambridge: Cambridge University Press.

    2004. "Labour, War and World Politics: Contemporary Dynamics in World-Historical Perspective." In: *Labour and New Social Movements in a Globalising World System*, edited by Marcel van der Linden and Christine Schindler Berthold Unfried. Vienna: Akademische Verlagsanstalt.

Simbulan, Roland G. 1985. *The Bases of Our Insecurity: A Study of the US Military Bases in the Philippines*. Metro Manila, Philippines: BALAI Fellowship.

Simpson, Dwight J. 1972. "Turkey: A Time of Troubles." *Current History* January:38–43.

Singer, P. W. 2003. *Corporate Warriors: The Rise of the Privatized Military Industry*. Ithaca, NY: Cornell University Press.

Smith, Neil. 2003. *American Empire: Roosevelt's Geographer and the Prelude to Globalization*. Berkeley: University of California Press.

Smyser, W. R. 1990. *Restive Partners: Washington and Bonn Diverge*. Boulder, CO: Westview Press.

Spoo, Eckhardt (ed.). 1989. *Die Amerikaner in der Bundesrepublik: Besatzungsmacht oder Buendnispartner?* Köln: Kiepenheuer & Witsch.

Steinke, Rudolf and Michel Vale (eds.). 1983. *Germany Debates Defense: The NATO Alliance at the Crossroads*. New York: M. E. Sharpe.

Steinweg, Reiner. 1982. "Die Bedeutung der Gewerkschaften für die Friedensbewegung." In *Die neue Friedensbewegung: Analysen aus der Friedensforschung*, edited by Reiner Steinweg. Frankfurt: Suhrkamp.

Stivers, William. 1986. *America's Confrontation with Revolutionary Change in the Middle East, 1948–83*. New York: St. Martin's Press.

Stolzfus, Saundra, Pollock Sturdevant, and Brenda Stoltzfus. 1992. *Let the Good Times Roll: Prostitution and the US Military in Asia*. New York: The New Press.

Suter, Lieutenant Colonel Thomas C. 1983. "Base Rights Agreements." *Air University Review* July–August.

Sweringen, Bryan van. 2008. "Stationing within the State: The U.S. Army Presence in the Rhineland-Palatinate 1947–2007." In *Amerikaner in Rheinland-Pfalz*. Wvt Wissenschaftlicher Verlag: Trier.

Szabo, Stephen F. 2004. *Parting Ways: The Crisis in German-American Relations.* Washington, DC: Brookings Institution Press.

Tarrow, Sidney. 1989. *Democracy and Disorder: Protest and Politics in Italy 1965–75.* Oxford: Clarendon Press.

Taşpınar, Ömer 2005. "The Anatomy of Anti-Americanism in Turkey." The Brookings Institute.

Thomas, Nick. 2003. *Protest Movements in 1960s West Germany: A Social History of Dissent and Democracy.* Oxford: Berg.

Thompson, Peter L. 1996. "United States-Turkey Military Relations: Treaties and Implications." *International Journal of Kurdish Studies* 9:103–113.

Tilly, Charles. 1984. *Big Structures, Large Processes, Huge Comparisons.* New York: Russell Sage Foundation.

   1985. "War Making and State Making as Organized Crime." In *Bringing the State Back In*, edited by P. B. Evans, D. Rueschemeyer, and T. Skocpol. Cambridge University Press, New York.

   1999. "From Interactions to Outcomes in Social Movements." In *How Social Movements Matter*, edited by Marco Guigni. Minneapolis: University of Minnesota.

Tilly, Charles and Sidney Tarrow. 2007. *Contentious Politics.* Boulder, CO: Paradigm Publishers.

Tolmein, Oliver. 1999. *RAF – Das war für uns Befreiung: Ein Gespräch mit Irmgard Möller über bewaffneten Kampf, Knast und die Linke.* Hamburg: Konkret Literatur Verlag.

Tuğal, Cihan. 2007. "NATO's Islamists: Hegemony and Americanization in Turkey." *New Left Review* 44:5–34.

Ulus, Özgür Mutlu. 2011. *The Army and the Radical Left in Turkey: Military Coups, Socialist Revolution, and Kemalism*, London: I.B. Tauris.

Ünüvar, Kerem. 2007. "Türk Solunun Türk Sağına Armağanı: 'Anti-emperyalizm.'" *Birikim* 214:106–112.

Uslu, Nasuh, Metin Toprak, Ibrahim Dalmis, and Ertan Aydin. 2005. "Turkish Public Opinion Towards the United States in the Context of the Ira Question" *The Middle East Review of International Affairs* 9:75–107.

Vali, Ferenc A. 1971. *Bridge Across the Bosporus: the Foreign Policy of Turkey.* Baltimore, MD: The Johns Hopkins Press.

Vasi, Ion Bogdan. 2006. "The New Anti-War Protests and Miscible Mobilizations." *Social Movement Studies* 5:137–153.

von Bredow, Wilfried and Rudolf H. Brocke. 1987. *Krise und Protest: Urspruenge und Elemente der Friedensbewegung in Westeuropa.* Opladen: Westdeutscher Verlag.

Walt, Stephen M. 2005. *Taming American Power: The Global Response to U.S. Primacy.* New York: W. W. Norton.

Wasmuth, Ulrike C. 1987. *Friedensbewegungen der 1980er Jahre: Zur Analyse ihrer strukturellen und aktuellen Entstehungsbedingungen in der BRD und den USA: Ein Vergleich.* Giessen: Focus

Wendt, Alexander. 1999. *Social Theory of International Politics.* New York: Cambridge University Press.

Wetzlaugk, Udo. 1988. *Die Allierten in Berlin.* Berlin: Berlin Verlag.

Wilkes, Owen. 1973. *Protest; Demonstrations against the American Military Presence in New Zealand: Omega 1968, Woodbourne 1970, Mount John 1972, Harewood/ Weedons 1973.* Wellington: A. Taylor.

Williams, Phil. 1985. *The Senate and US Troops in Europe.* New York: St. Martin's Press.

Willoughby, John. 2001. *Remaking the Conquering Heroes: The Social and Geopolitical Impact of the Post-War American Occupation of Germany.* New York: Palgrave.

Wittner, Lawrence. 2003. *Toward Nuclear Abolition: A History of the World Nuclear Disarmament Movement.* Stanford: Stanford University Press.

Wolf, Charlotte. 1969. *Garrison Community; A Study of an Overseas American Military Colony.* Westport, CT: Greenwood Pub. Corp.

Woodliffe, John. 1992. *The Peacetime Use of Foreign Military Installations under Modern International Law.* Boston: Dordrecht.

Yeremenko, Marshal A. 1960. "Arguments Against Foreign Bases." *International Affairs*, published in Moscow November 1960.

Yetkin, Murat. 2004. *Irak Krizinin Gerçek Öyküsü.* Istanbul: Remzi Kitabevi.

Zepp, Marianne. 2007. "Redefining Germany: Reeducation, Staatsbürgerschaft und Frauenpolitik Im US-Amerikanisch Besetzten Nachkriegsdeutschland." Osnabrück: V&R Unipress

Zimmermann, Hubert. 2002. *Money and Security: Troops, Monetary Policy, and West Germay's Relations with the US and Britain, 1950–1971.* Cambridge: Cambridge University Press.

Zirakzadeh, Cyrus Ernesto. 2006. *Social Movements in Politics: A Comparative Study.* New York: Palgrave.

Zorns, Redaktionsgruppe Früchte des (ed.). *Die Früchte des Zorns: Texte und Materialien zur Geschichte der Revolutionären Zellen und der Roten Zora.* Amsterdam: Edition ID-Archiv.

    (ed.). 1993. *Die Früchte des Zorns: Texte und Materialien zur Geschichte der Revolutionären Zellen und der Roten Zora.* Amsterdam: Edition ID-Archiv.

Zürcher, Erik J. 2003. *Turkey: A Modern History.* London: I. B. Tauris & Co.

Newspapers and Other Online News Sources

Bağımsız iletişim ağı
Chicago Tribune
Christian Science Monitor
Financial Times
Frankfurter Allgemeine Zeitung
Frankfurter Rundschau
Hürriyet
Le Monde diplomatique
Los Angeles Times
Mideast Mirror
Milliyet
The New York Times
Die Tageszeitung
Turkish Daily News
The Wall Street Journal
The Washington Post
Zaman
Die Zeit

Index